Justice for Future Generations

For Anja,
Hannah, Sylvia, Emma (2001–2010) and Margaret

Justice for Future Generations

Climate Change and International Law

Peter Lawrence

Senior Lecturer, Faculty of Law, University of Tasmania, Australia

Edward Elgar

Cheltenham, UK • Northampton, MA, USA

Published by
Edward Elgar Publishing Limited
The Lypiatts
15 Lansdown Road
Cheltenham
Glos GL50 2JA
UK

Edward Elgar Publishing, Inc.
William Pratt House
9 Dewey Court
Northampton
Massachusetts 01060
USA

A catalogue record for this book
is available from the British Library

Library of Congress Control Number: 2013957762

This book is available electronically in the ElgarOnline.com
Law Subject Collection, E-ISBN 978 0 85793 416 1

FSC
www.fsc.org
MIX
Paper from
responsible sources
FSC® C013056

ISBN 978 0 85793 415 4

Typeset by Columns Design XML Ltd, Reading
Printed and bound in Great Britain by T.J. International Ltd, Padstow

Contents in brief

Full contents

Preface

SUMMARY OF THE ARGUMENT OF THIS BOOK

Climate change raises acute issues of justice between generations. This is because the worst climate change impacts will be experienced by future generations who did not create the problem. These impacts include extreme weather events and droughts and even the disappearance of some small island states, with the poor being particularly affected. Climate change science makes clear that without urgent action being taken now to sharply reduce greenhouse gas (GHG) emissions, there is a strong likelihood of harm to the interests of future generations and indeed a significant risk of catastrophic harm. Moreover, delays by the current generation increase the costs of mitigating and adapting to climate change for our descendants (Chapter 1).

In this book, I identify a number of criteria – justice principles – essential for ensuring intergenerational justice in relation to the mitigation of climate change. These principles are used to address the overarching question of this book: *what does justice require of current generations in addressing climate change to safeguard the welfare of future generations and how should such obligations be reflected in international law?*

The justice principles developed in this book are applied to GHG emissions conceived as a limited 'resource'. Climate mitigation involves the question of how to fairly distribute a limited allocation of 1 trillion tonnes of CO_2 between now and 2050 (Shue 2011). This is a daunting task given ever-increasing energy demand. It is also a task which needs to be addressed in a multidisciplinary way, drawing on, for example, insights from international relations theory as to the role of justice in treaty negotiations, and political science, in explaining why the current international climate change regime reflects so poorly the interests of future generations.

Part I of this book ('Theory') draws on theories of justice and ethics to address the question of what ethical obligation contemporaries owe to future generations in relation to climate change mitigation and how the mitigation burden should be distributed between contemporaries and future generations. An ethical obligation towards future generations in this context is presented as resting on a harm avoidance principle and the notion of future generations possessing human rights (Chapter 2). Deep cuts in GHG emissions are required in order to avoid harm to the interests of future generations and to preserve their core human rights. A premise in this approach is that core human rights to life, subsistence and health are entitlements that all persons possess *equally* regardless of when and where they live. Both these core rights and respect for human dignity have received virtually universal recognition in the Universal Declaration on Human Rights (UDHR) and key human rights instruments. Moreover, there is widespread support for these core human rights and the notion of 'human dignity' from diverse belief systems including the major world religions.

An advantage of this approach is that it does not rely upon one single ontological moral basis for human rights. The argument takes as a starting point the widespread support for the values expressed in these core human rights. By using a lean list of rights it is unnecessary to delve into the issue of whether other rights, such as rights to democratic participation or freedom of expression, are shared values or, rather, inevitably linked to Western political liberalism (Caney 2010a).

This book relies on anthropocentric 'social justice' theories rather than deep ecology approaches that ascribe intrinsic value to nature for two reasons. First, anthropocentic approaches are more likely to gain political traction and help form a basis for effective climate change mitigation. Second, deep ecology theories generally point in the same direction as the social justice theories relied upon here (Chapter 2).

One such social justice theory is the 'capabilities approach' of Sen and Nussbaum, which provides a good basis for an obligation towards future generations as a failure to take mitigation action will impair the ability of future generations to fulfil their core capabilities. A 'communitarian' approach points in the same direction on the basis of a transgenerational community extending into the future. At first blush communitarian theories seem to be unable to provide a basis for obligations owed to non-compatriots. But this limitation can be overcome by 'cosmopolitan' theories which provide a basis for obligations across national boundaries, achieved in a global community in relation to core human rights impacted by climate change and the interdependence of all people relating to climate change causes and impacts (Chapter 2).

The notion of core human rights essential for human dignity and a harm avoidance principle provide the basis for an ethical obligation on governments to take strong mitigation action to ensure that the global climate system does not pass beyond a threshold which results in significant harm to the fundamental interests and core human rights of future generations. But this ethical obligation remains incomplete. Principles of distributive justice are required, as well as clarity, in terms of which states bear the obligation (for example, only developed?), the relevance of historic emissions and knowledge of damage. This book proposes a number of justice principles suitable for determining how the mitigation burden should be fairly distributed between current and future generations, including the rate at which mitigation should occur (Chapter 3). Achieving justice for future generations is dependent on substantially meeting these distributional justice principles. Three key principles are identified: a *core human rights* principle, *avoidance of harm* principle and *capacity to pay* principle. These principles are combined with a *sufficiency* notion of justice which places priority on achieving a basic subsistence level for as many people as possible, and an *equality* notion which seeks to equalise opportunity or outcomes (Page 2006: 79–80).

In justifying these principles I draw on various theories of justice and ethics. These theories are not justified – a task beyond the scope of this book – but are taken as starting points. A number of the principles, however, already have widespread support in the international system. Thus, the *sufficiency* notion of justice is recognised by the international community as a widely shared value in, for example, the UN Millennium Goals. *Equal* entitlement to core human rights is also recognised by the international community in, for example, the Universal Declaration of Human Rights. *Sufficiency* and *equality* notions of justice are argued as requiring the apportionment of the global emissions 'budget' between now and 2050 on the basis of (1) an equal per capita emissions basis, modified to take into account national circumstances (3.5.9), and (2) the notion that the poor should be required to reduce emissions last owing to their sufficiency requirements (3.5.10). These later two principles are amongst five 'Implementation Principles' (IPs) which flow from the key principles and overarching core principles of *equality* and *subsistence*.

Assuring a reasonable balance between the needs of the contemporary poor and future generations is a particularly challenging issue given the reality that rapid movement to a low-carbon economy in developing countries will involve considerable economic and social dislocation. To address this concern a 'structural reform principle' is proposed which places an onus on policy-makers to not delay structural reforms necessary to ensure the protection of the long-term interests of future generations while avoiding harm to the core rights of current generations (3.5.3).

To ensure intergenerational justice, the justice principles must operate under an *effectiveness imperative* based on Article 2 of the United

Nations Framework Convention on Climate Change (UNFCCC) and involving the prevention of dangerous anthropocentric interference with the climate system (3.4.1). This effectiveness imperative must be met in order to avoid harm to future generations. The *effectiveness imperative* depends upon duties of precaution and cooperation. While the latter already find embodiment in the UNFCCC, the ethical basis of these principles remains important in order to demonstrate their deeper legitimacy in justice and ethics. This deeper legitimacy is integral to the approach of this book in elaborating a justice-based normative framework used to evaluate international law rules.

Thus Part II of this book ('International Law and Politics') evaluates current international law rules on climate change in order to assess, firstly, the extent to which the justice principles, implementation principles and *effectiveness* imperative are met, and, secondly, to explain the poor embodiment of justice principles in current international law rules. The methodology adopted in Chapter 4, which addresses the first issue, rests on an approach to sources of international law according to which state consent is central, while acknowledging that so-called 'soft-law' norms may have considerable impact on national law and in shaping expectations. In order to meet the *effectiveness imperative*, scientists are calling for a virtual decarbonisation of the global economy by 2050 with 10–40 per cent GHG emission reductions by 2020 (Chapter 1). Such emission reductions need to be anchored in a global treaty which includes binding mitigation targets, a funding mechanism to facilitate technology transfer and assist developing countries in reducing GHG emissions and an effective compliance mechanism (Chapter 4). The current global climate treaty regime falls well short of these requirements. The Kyoto Protocol has been further extended to cover 2013–20 but only imposes GHG emission reduction targets on the EU and a few other industrialised countries.

Under the Durban platform agreed at COP17 in 2011, there is a mandate to negotiate a global agreement by 2015 to enter into force by 2020. This timetable fails to meet the *effectiveness imperative* as climate scientists are calling for GHG emissions to peak between 2015 and 2020 and rapidly decline thereafter (Allison et al. 2009: 7). Moreover, the delay in making emissions entailed in this time frame would mean that emissions would have to be reduced more sharply than otherwise, which may not be feasible (Macintosh 2010). Furthermore, there is ambiguity under the Durban platform as to whether the instrument to be negotiated must be a treaty status agreement (Chapter 4). Absent binding mitigation targets, covering all industrialised countries and major developing country emitters, the existing compliance mechanism and funding mechanisms remain inadequate, like the walls of a house without foundations.

The principle of 'intergenerational equity' is examined in general international law in order to see if this sheds light on how this principle should be interpreted in its embodiment in Article 3 of the UNFCCC. This analysis also aims to establish whether this principle – alone or as an element of sustainable development – can be relied upon to establish obligations which meet the justice principles of Chapter 3, independently of the UNFCCC. 'Intergenerational equity' is shown to lack a consistent formation in global environmental treaties sufficient for it to be a binding norm of customary international law. Moreover, this concept is argued to be indeterminate and incapable in itself of addressing the substantive justice requirement of prescribing what weight should be given to future generations' interests vis-à-vis current generations.

A survey of jurisprudence of the International Court of Justice (ICJ) shows that intergenerational equity as an element of sustainable development has played a role in combination with other international law rules. However, again the inherent indeterminacy of 'sustainable development' means that it is incapable of assisting in allocating climate change mitigation burdens which can only be effectively achieved by application of justice principles in rules embedded in a global treaty. This is not to deny the importance of sustainable development which is argued to be essential in implementing obligations owed to future generations in relation to climate change.

Unmitigated climate change threatens the human rights to life, health and subsistence enshrined in UN human rights instruments (Chapter 5). However, litigation based on violations of international human rights rules relating to climate-change-related harms faces obstacles in terms of establishing standing, causation and responsibility for extraterritorial harms – namely, harms that occur outside the state that is responsible for breaching the particular obligation. These challenges are even more acute in relation to future generations, where harm has not yet occurred. Human rights litigation is explored in both the EU and US, where courts have been reluctant to interfere in climate policy-making, considered to be within the political domain.

Human rights has been proposed as an appropriate basis for benchmarks from which mitigation targets may be derived. An advantage of using human rights in identifying benchmarks in relation to climate change mitigation rests in wide acceptance of core human rights. However, a limitation of human rights is that it cannot address the distributional justice issues involved in intergenerational justice including, for example, how rapidly GHG emissions should be reduced.

This book breaks new ground by using 'discourse theory' to explain why the international climate regime so weakly embodies intergenerational justice (Chapter 6). 'Discourses', following Dryzek (1997) and Hajer (1995), comprise shared understandings. Dominant discourses include 'industrialism', which seeks to maximise goods and services, 'ecological modernisation', which requires environmental harm to be internalised in the costs of goods and services, and 'climate marketisation', which dictates the use of market mechanisms such as emissions trading schemes. This book builds on the work of Dryzek and Stevenson (2011) in analysing recent climate change negotiations. These dominant discourses have been underpinned by powerful economic interests of fossil fuel and related high GHG-dependent industries and lobbying networks. While the discourse of *ecological modernisation* incorporates notions of intergenerational justice, the rhetoric has not been matched with action, with industrialised countries generally failing to adequately reduce GHG emissions. Importantly, developing country statements in the UN climate negotiations have attributed to industrialised countries a *global responsibility* for ensuring protection of future generations from climate change. A shift in these discourses and the economic interests that underpin them is essential for progress in climate change mitigation.

Part III of this book explores how justice principles *should* be incorporated in international law rules relating to climate change. I argue that a notion of global responsibility needs to be fostered in order to reflect the notion of proportionate responsibility rather than industrialised countries being seen as having sole responsibility for future generations (Chapter 7). This will, however, only occur with political leadership in the North. Industrialised countries should foster trust by addressing the development concerns of developing countries in other forums such as the World Trade Organization (WTO) (Roberts and Parks 2007). Moreover, industrialised countries need to sharply reduce emissions to build trust inside the UNFCCC regime.

International law, particularly in the form of treaty rules, remains crucial in ensuring intergenerational justice in relation to climate change. An agreement binding under international law is more likely to give the stability of commitments required to meet long-term objectives when compared with other non-binding options such as the Copenhagen pledge and review model (Chapter 7). A treaty-level instrument is also essential to address trade competitiveness concerns. The prevailing consensus approach to treaty-making in the UN hinders rapid development of an effective climate treaty. However, with political leadership, and a shift in the dominant discourses, discussed below, these processes can potentially deliver an effective climate regime (Chapter 7).

While not central to this book, some mechanisms are canvassed aimed at promoting procedural climate justice, including the idea of an international commissioner for future generations and reform of UNFCCC procedural rules to allow youth climate NGOs greater opportunities to participate in the negotiation process on the basis that they are more likely to represent the interests of future generations (7.6.1). It remains unclear whether such procedural mechanisms would be successful in delivering substantive justice and further research in this area is required.

National proposals for a future global climate agreement are evaluated against the justice principles and *effectiveness imperative* set out in Chapter 3. This evaluation also uses 'political feasibility' criteria (following Bosetti and Frankel 2011), which require a regime to include all major emitters and not cost any individual party more than a certain threshold. Following Pickering et al. (2012), an 'institutional continuity' criteria is used whereby proposals must build on the existing architecture and principles of the UNFCCC. This book does not include a full analysis of political feasibility; the more modest aim is to examine the feasibility of substantially meeting the requirements of the justice principles and *effectiveness imperative* (Chapter 3) in a global climate treaty.

It is too early in the Durban platform negotiation process for fully-fledged proposals to emerge. However, skeletal proposals made by some countries, and proposals made under the Bali mandate, which may well influence future proposals, are explored (Chapter 7). Proposals by China, India and Brazil reflect strongly the principles of *responsibility for harm* and *capacity to pay* by placing mitigation targets exclusively on industrialised countries. However, these proposals are problematic in terms of addressing the *effectiveness imperative* and *political feasibility* criteria, both of which require legally binding commitments on major developing countries. A proposal by the EU is stronger in terms of the *effectiveness* and the *responsibility for harm* principle. The EU is committed to a 20 per cent reduction in GHGs by 2020 and will raise this to 30 per cent but only if other major emitters make comparable emission reductions.

Absent from the national proposals are some essential elements required for a global climate treaty to achieve intergenerational justice. While the UNFCCC contains a reference to sustainable development, this falls short of the required commitment to strong sustainability. The latter means that elements of the global ecosystem which are not substitutable – including the ongoing stability of the global climate – are not traded off against other elements. Also vital is a commitment to a *structural reform implementation principle* which places an onus on policy-makers not to delay the structural reforms necessary to ensure the protection of the long-term interests of future generations while minimising harm to the core rights of current generations.

While these principles do not necessarily need to be embedded in a new climate treaty, a shared understanding of them is a prerequisite for progress. Indeed, international relations literature suggests that no single notion of fairness needs to be accepted by all as a prerequisite for agreement (Albin 2001). A compromise involving different competing justice elements may be possible. While the exact elements of such a treaty would need to be negotiated, the broad contours of an instrument which would meet both the requirements of intergenerational justice and political feasibility are identified (7.5.1). A treaty of this nature would need to include both short- and long-term national emission reduction targets quantified within an overall budget requiring zero emissions by 2050. Such a treaty could meet the requirements of political feasibility by placing a cost cap on emission reductions to be taken each year by each country. In the short term it would reflect a grandfathering approach combined with capacity to pay, while – over the long term – moving towards equal per capita emissions (Bosetti and Frankel 2011). Larger developing countries would take on emissions gradually over time, reflecting the principle of *proportionate responsibility for harm*. This would depend on developed countries making sharp emission reductions and substantially funding developing country emission reductions. Historic emissions would be reflected in the treaty's funding mechanism rather than in the framing of mitigation targets given difficulties with measuring historic emissions (Pickering et al. 2012) (7.5.1). While human rights provide a strong ethical basis for mitigation action to protect future generations, an injection of human rights discourse into the climate change negotiations could exacerbate existing North–South tensions. Thematic rapporteurs within the UN human rights system (for example on the right to food) have ensured increased and welcome attention to climate change impacts on the enjoyment of particular human rights. However, the creation of a special rapporteur on climate change is argued to be of limited value (Chapter 7).

The upshot of this analysis is that it is difficult but possible to implement justice principles to safeguard the welfare of future generations in a workable global treaty. But this will only occur with a shift in the dominant discourses and economic interests which underpin them. As a minimum, there needs to be urgent internalisation of climate change harms through effective regulation. In addition, ethics and justice-based discourses need to become more prominent in climate policy-making. High discount rates applied by some economists are problematic as they imply that future persons do not have equal value with contemporaries (1.8). While economic modelling is important in ensuring that governments make informed policy choices in terms of climate change mitigation, justice- and ethics-based assumptions need to be transparent and based on shared values. Crafting strong international law rules to address climate change remains a vital task for humanity and the global ecology. Justice for future generations depends upon it. Building agreement on what justice means in this context is an essential part of this task.

Acknowledgements

I must thank Jonathan Vershuuren (University of Tilburg Faculty of Law) for his constant support, wisdom and expertise throughout this project. Invaluable comments were also received from C.J. Bastmeijer, W.J.M. van Genugten, H. Lindahl (all of University of Tilburg Faculty of Law) and L. Kotzé (North West University).

Support from the University of Tasmania (UTAS) Law School, especially my Dean, Margaret Otlowski, and my mentor, Gino Dal Pont, has been constant and generous. The library staff, particularly Deb Bowring, met all my research needs with speed and good humour.

Jonathan Pickering, Tim Stephens and my father John Lawrence provided invaluable feedback in the development of this book.

Nathan Bindoff, an internationally renowned climate scientist, provided generously of his time in reviewing the science sections of Chapter 1 of this book. Nathan is coordinating lead author for the Oceans chapter in the Intergovernmental Panel on Climate Change (IPCC)'s Fourth Assessment Report and a coordinating lead author for the Fifth Assessment Report under preparation.

Hayley Stevenson generously provided transcripts of UN climate change meetings – and helpful feedback – essential for the argument on climate change discourses (Chapter 6).

Various parts of this book were presented at a number of institutions and I would like to thank the participants and members of the audience in each case: London School of Economics and Political Science (LSE) Philosophy and Public Policy Lecture Series (organised by Richard Bradley and Luc Bovens), October 2008; Australian National University (ANU), Canberra, August 2009, Workshop on Environmental Discourses in Public and International Law (organised by Brad Jessup and Kim Rubenstein); ANU College of Law staff seminar, Canberra, March 2011; Wilhelm-Merton-Centre for European Law and International Law, Goethe University of Frankfurt, May 2011; Smith Centre for Enterprise and the Environment, University of Oxford, May 2011 (hosted by Chukwumerije Okereke, then Director of the Smith Centre Oxford); and the Australian and New Zealand Society of International Law (ANZSIL), Canberra, June 2011.

Michael Bothe and Rainer Hofmann generously hosted my visit to the Wilhelm-Merton-Centre, University of Frankfurt in May 2011. Richard Kinley and Daniel Klein of the UNFCCC Secretariat in Bonn kindly hosted a productive visit to the Secretariat in May 2011. Harro van Asselt facilitated my visit to Oxford in May 2011 and helped get this book started.

UTAS generously provided me with study leave for the first half of 2011, which included financial assistance, to make these visits possible.

I must thank the ANU College of Law for a Visiting Fellowship in February 2011. Kim Rubenstein, then Director of the Centre for International and Public Law (CIPL), was a generous host. Staff at the Australian Centre for Environmental Law (ACEL) directed by Tim Bonyhady provided invaluable assistance, especially Andrew McIntosh.

The following provided helpful feedback on drafts or assisted in clarifying my views: Harro van Asselt, Bridget Dunne, Sam Fankhauser, Matthias Hartwig, Marcus Hayward, Brad Jessup, Orr Karassin, Matt Killingsworth, Robert Kirby, Daniel Klein, Jan McDonald, Jeff McGee, Geoff Robin, Shirley Scott, Rick Snell, Fredrich Soltau, Hayley Stevenson, Michael Stokes, Greg Terrill, Joerg Tremmel, Steve Vanderheiden and Steve Waight.

Apologies if I have inadvertently omitted anyone who should be on this list.

Aaron Moss, Meg Good, Natalie Gillon, Katie Purchas and Laura Keys provided invaluable assistance with the bibliography and technical aspects of the manuscript.

Chapter 6 of this book draws upon material from within Peter Lawrence, 'Justice for future generations: environment discourses, international law and climate change', in Brad Jessup and Kim Rubinstein (eds), *Environmental Discourses in Public and International Law* (2012), © Cambridge University Press, reproduced with permission.

Ben Booth and his team at Edward Elgar have been a delight to work with.

My partner Anja made this book possible in too many ways to list here, including finding my Ph.D. supervisor and taking over my Jessup coaching duties in late 2012.

To my daughters: you are the inspiration for this book.

Finally I must thank Noisy Bob – my 10-year-old's pet rooster who got me up in time to ensure good progress towards the difficult final stages.

All errors and omissions are, of course, my responsibility.

The law is current as of 1 September 2013.

Abbreviations

AOSIS	Association of Small Island States
AWG-LCA	Ad Hoc Working Group on Long-Term Cooperative Action
BAU emissions	business-as-usual emissions
CBDR	common but differentiated responsibilities
CDM	Clean Development Mechanism
COP	conference of parties
EASD	equitable access to sustainable development
ECHR	European Convention on Human Rights
EIG	Environment Integrity Group
EU	European Union
FAO	Food and Agriculture Organization
GDP	gross domestic product
GHG emissions	greenhouse gas emissions
G77	Group of 77
ICCPR	International Covenant on Civil and Political Rights
ICESCR	International Covenant on Economic, Social and Cultural Rights
ICJ	International Court of Justice
ILA	International Law Association
IPCC	Intergovernmental Panel on Climate Change
LDCs	least developed countries
NGOs	non-government organisations
PPP	polluter pays principle
SRES	Special Report on Emissions Scenarios
UNDP	United Nations Development Programme
UNEP	United Nations Environment Programme
UNESCO	United Nations Educational, Scientific and Cultural Organization
UNFCCC	United Nations Framework Convention on Climate Change
USEPA	United States Environmental Protection Agency
WMO	World Meteorological Organization
WTO	World Trade Organization

Table of cases

International Court of Justice and Permanent Court of Arbitration

European Commission and Court of Human Rights

United States

Philippines

Australia

Table of international instruments

1. Introduction: the climate change problem and solutions

This book examines what justice requires of current generations in addressing climate change to safeguard the welfare of future generations and how such obligations should be reflected in international law. The first part of this question is addressed by identifying several essential criteria for ensuring intergenerational justice in relation to climate change. These criteria – or justice principles – provide a basis for critically assessing the existing international law regime for climate change. They also provide a springboard for reforming contemporary international climate change law.

The challenge of addressing anthropogenic climate change raises intergenerational equity or justice[1] issues. Future generations who did not cause climate change will be the most severely impacted. This applies not only to individual or generations' contributions to creating the problem, but also in relation to nation states. The most graphic example is that the small island states of Kiribati and Tuvalu, with negligible contribution to global greenhouse gas emissions (GHG), will be amongst the nations most severely affected, with the land mass of these Pacific island states becoming uninhabitable because of salt water intrusion and ultimately because they are completely submerged in the coming century.[2]

Climate change also raises issues of intragenerational justice in terms of which states and societies should bear the costs of mitigating climate change by reducing (GHG) emissions. The preponderance of scientific evidence indicates that we need a total decarbonisation of the global economy by 2050 – the elimination of GHG emissions from human activities – to have a reasonable chance of avoiding dangerous anthropogenic climate change (see 1.4 below). But we are currently locked into fossil-fuel-dependent economic development. To achieve these emission reductions it is apparent that substantial cuts in GHG emissions will be needed, by both industrialised countries *and* also at least the larger developing countries (Gupta et al. 2007: 751).[3] However, developing countries[4] have resisted binding targets in the UN negotiations, arguing that the industrialised countries created the problem through their historic emissions and that only when industrialised countries have made deep cuts will developing countries take action. Developing countries have been emphasising justice towards current generations as being of paramount importance, with a strong emphasis on the need to address poverty now (Scholtz 2009: 167).[5] Industrialised countries have emphasised in their rhetoric the need to take action now for the benefit of future generations (Scholtz 2009: 170) although to date most have done relatively little in terms of curbing GHG emissions (UNFCCC 2012). The international negotiations on climate change may accordingly be characterised as both a conflict between justice and effectiveness,[6] and a conflict between intergenerational equity and intragenerational equity.

Curiously, while there is a wealth of literature by philosophers and public policy experts on intergenerational justice, there is very little by international legal scholars.[7] This is despite the fact that there are strong interconnections between philosophical models of intergenerational equity and international law formulations of the same concept. A central argument in this book is that some of the philosophers' insights in relation to ethics and justice are of great value in analysing the international law principle of intergenerational equity. As Koskenniemi (2005) has argued, international law performs a dual function in justifying state action but also as a basis for critiquing state action through its incorporation of ethical principles. We shall see this dual function is very much in evidence when we explore the ethical principles relating to intergenerational justice and their – albeit weak – incorporation in international law rules. Before examining the justice and international law dimensions of these issues, it is essential to have an understanding of the salient aspects of climate change science.

This chapter sets out the key elements of climate change science (1.1), including the notions of abrupt climate change and tipping points (1.1.2). This leads to a discussion of climate change impacts (1.1.3), including the key concepts of risk and uncertainty (1.1.4). The global mitigation response required to address the climate change challenge is addressed with a focus on what is known as a 'budget' approach which identifies the volume of carbon that can safely be stored in the atmosphere before it causes dangerous climate change (1.2). The discussion then moves to the justice and ethics issues raised by the science (1.3), with the distinction drawn between intragenerational and historic justice (1.4) and intergenerational justice (1.5). Procedural justice issues are addressed (1.6) before moving on to discuss the notion of 'discounting' employed by economists (1.8). The focus on international law in this book is explained (1.9) in the context of the current climate change negotiations and the existing UN climate change regime (1.10). Conclusions are drawn in 1.11.

1.1 CLIMATE CHANGE SCIENCE

Climate change refers to 'change in the average pattern of weather over a long period "typically decades or longer"' (Australian Academy of Science 2010: 4). The basic processes of climate change are well understood. Since the late nineteenth century it was known that increases of carbon dioxide (CO_2) into the atmosphere 'would act like added insulation in the earth's atmosphere, trapping more heat near the surface' (Australian Academy of Science 2010: 10). GHG emissions play a key role in determining climate and bringing about climate change. 'Greenhouse gases include water vapour, carbon dioxide (CO_2), methane, nitrous oxide and industrial gases such as chlorofluorocarbons (CFCs)' (Australian Academy of Science 2010: 4). These gases function as an 'insulating blanket', maintaining the earth's surface temperature (Australian Academy of Science 2010: 4). However, with the exception of water vapour, the concentrations of all of these gases in the atmosphere are being influenced by human activities. Concentrations of these gases, particularly CO_2, have increased dramatically since the Industrial Revolution, particularly as a result of fossil-fuel-burning and other industrial processes including cement production, but also from deforestation linked to agriculture (Australian Academy of Science 2010: 10). Increasing population has been an integral aspect of this process and will continue to be so into the future (IPCC 2007a: 18).

A particular feature of CO_2 – the most important GHG – is vital in terms of the intergenerational issues considered in this book. This is the persistence of CO_2 in the atmosphere. Scientists have now discovered that, unlike trace gases with well-defined timelines, CO_2 persists in the atmosphere for a lengthy period of time. The impact of this physical property of CO_2 is dramatic:

> Stabilization of CO_2 emissions at current levels will result in a continuous increase of atmospheric CO_2 over the 21st century and beyond ... In fact, only in the case of essentially complete elimination of emissions can the atmospheric concentration of CO_2 ultimately be stabilized at a constant level (Meehl et al. 2007: 824–5).

Put differently, even if today human beings ceased activities causing GHG emissions entirely, concentrations of CO_2 in the atmosphere would continue with only a reduced rate of growth over the coming decades (Meehl et al. 2007: 824–5).

Climate change science became known to a wider audience in 1988 when the Intergovernmental Panel on Climate Change (IPCC) was created by the United Nations Environment Programme (UNEP) and the World Meterological Organization (WMO). The IPCC was tasked to assess in an 'objective, open and transparent' way the scientific, technical and socio-economic information relevant to understanding human-induced climate change while remaining 'neutral with respect to policy' (IPCC 2006: para 2). The IPCC carries out this assessment by bringing together scientists from around the world and seeking to reach agreement on various aspects of climate change, from the physical causes and impacts to mitigation and adaptation issues (IPCC 2006: para 2).

The most recent report of the IPCC is the Fourth Assessment Report, released in 2007 (IPCC 2007a), which was 'written by more than 450 lead authors, 800 contributing authors and reviewed by over 2,500 expert and government reviewers', with many of the lead authors considered the world's top experts in their field (Anderegg and Harold 2009: 7). The IPCC assessment reports can reasonably be interpreted as the state of the mainstream science on climate change (Anderegg and Harold 2009: 4). Steffen (2009: 4) points out that anthropogenic emissions of CO_2 have been 'rising at or near the upper limit of the envelope of the IPCC projections since they were first published in 1990'. While 'considerable uncertainty' remains in relation to estimates of future sea-level rise, empirical observations have confirmed that actual levels of sea-level rise have been occurring 'toward the upper range of the IPCC projections' (Steffen 2009: 11).

The Fourth Assessment Report of the IPCC concluded with 'very high confidence' – defined as more than 90 per cent likelihood – that 'the net effect of human activity since 1750 has been one of warming' (IPCC 2007a: section 1). It went on to conclude that 'most of the observed increase in global average temperatures since the mid 20th century is *very likely* due to the observed GHG concentrations' (IPCC 2007a: section 1). The Report concluded that fossil fuel emissions were the most significant source of GHG emissions with agriculture-related emissions also highly significant. Since the IPCC Fourth Assessment Report, the assessment of attribution has been further strengthened in the scientific literature (Stott et al. 2010).[8]

The IPCC Fourth Assessment Report has used a range of models to make projections as to global average surface warming into the future based on a range of assumptions relating to technological change and economic and population growth. These projections are in the range of 1.1–6.4°C (with the best estimate of 1.8–4°C) by the end of the twenty-first century (IPCC 2007b: 13).

There is overwhelming support amongst the mainstream – peer-reviewed – scientific community in relation to the key aspects of climate change science. This is confirmed by reports of numerous national academies of science (Garnaut 2011: 47). If a consensus is defined as more than 95 per cent agreement, surveys of climate scientists' peer-reviewed research outputs have concluded that there is a consensus (around 97 per cent) agreement in relation to the key propositions. These propositions are that global average temperature has increased since the preindustrial times and that anthropogenic GHG emissions are a significant contributing factor in these temperature increases (Anderegg and Harold 2009: 15).[9] However, significant uncertainty remains as to the rate and regional distribution of climate change impacts, including for example changes in rainfall distribution (Stott et al. 2010).

A small minority of scientists continue to call into question the mainstream science, but a substantial number of these are not qualified scientists or have geology – not climate science – qualifications (Anderegg 2010: 31). The views of those who challenge the mainstream science – 'deniers', often mislabelled as 'sceptics' (all scientists are sceptics) – have gained widespread publicity, partly owing to deliberate misinformation campaigns by powerful vested interests (this is explored in Chapter 6). This book takes as its starting point the mainstream science.

1.1.1 Abrupt Climate Change and Tipping Points

Projections of future climate change impacts have involved a *gradualist* paradigm according to which climate change outcomes are largely an extension of projections based on current processes. In addition, more attention is gradually focusing on an *abrupt* paradigm involving the crossing of particular thresholds with catastrophic impacts (Gardiner 2009: 141–2). The abrupt paradigm is linked to so-called 'tipping points'. A tipping point is a critical threshold at which the future state of a system can be qualitatively altered by a small change (Lenton et al. 2008). Abrupt climate change refers to tipping point change 'which occurs faster than its cause' (Allison et al. 2009: 40).

The most well-known potential tipping points include the melting of the Greenland ice sheet, and the West Antarctic ice sheet, both with the potential for 'large and global sea level rise' (Lenton et al. 2008, Kriegler et al. 2009). A further tipping point relates to dieback of the Amazon rainforest, where it has been suggested that there is a 2°C temperature increase threshold. 'Beyond this threshold the area of dieback rises rapidly from over 20 to 60 percent' (Jones and Lowe 2011).[10]

Kriegler et al. (2009) have conducted a survey of 43 experts to seek to gain a better understanding of the probabilities of various tipping points occurring. The conclusion of this study was that while there was large uncertainty among experts about the prospects of triggering major changes in the climate system, this did not imply that the probability of such outcomes was remote (Kriegler et al. 2009: 7). In fact, the study found that the experts allocated 'significant probability' to some tipping point events, such as the dieback of the Amazon rainforest and melting of the Greenland ice sheet (Kriegler et al. 2009: 1).

1.1.2 Impacts

Certain climate change impacts predicted by the IPCC will particularly impact on future generations. Impacts predicted by the IPCC to be likely or very likely include the following:

- rising sea levels
- human health impacts including increases in mortality as a result of increased frequency of extreme weather events, including floods, storms and heatwaves
- the increased transmission area of various disease agents
- increased droughts
- increased extinction of various species
- the permanent acidification of the oceans (IPCC 2007c: 9–13).

A number of these impacts are permanent and irreversible, for example species extinctions and ocean acidification. Moreover, some research since 2007 indicates that there may be a risk that aspects of the climate system enter into a state from which they cannot return (for example Jones and Lowe 2011).[11]

Impacts generally are predicted to be more severe for developing countries, particularly those in Africa and the 'Asian and African mega-deltas, due to large populations and high exposure to sea level rise, storm surges and river flooding' (IPCC 2007a: section 3.3.3). It is also important to recognise that there will be significant differences between the impacts of climate change within developing countries, with 'the poor, elderly and very young particularly impacted' (IPCC 2007a: section 3.3.3).

According to the most recent predictions, a number of the impacts sketched above will already be in evidence by 2020, impacting the lives of younger people alive today. For example, while the impact of climate change on water resources across the African continent is uneven, on some assessments the population 'at risk of increased water stress in Africa … is projected to be 75–250 million … by the 2020s' with reduced agricultural production and consequent impacts on poverty and food security (IPCC 2008: 81).

Edward Page (2006: 36) points out that '[t]he most vulnerable of all to climate change impacts will be *future* members of developing countries', and that the greatest inequalities in terms of climate change impacts will be between the current elites in industrialised countries and the future poorest members of developing countries. Further into the future, however, it remains less clear whether industrialised countries maintain their current status.

In conclusion, we can say that the mainstream view among scientists is that without deep GHG emission cuts there is a high likelihood of *severe impacts* resulting from climate change. We can also say that without deep GHG emission cuts there is a *substantial risk* of abrupt climate change involving catastrophic impacts.

1.1.3 Risk and Uncertainty

The propositions set out in the previous paragraph included the notions of 'likelihood' and 'risk'. Uncertainty is important in considering intergenerational justice and climate change for a number of reasons. Firstly, some have argued that the fact that scientists cannot predict with accuracy climate change impacts, for example at the regional level, is a justification for not taking mitigation action. According to this approach, mitigation action will *definitely* cause economic harm now, and this is not outweighed by future *uncertain* benefits of mitigation. This approach, however, is unconvincing and contravenes the precautionary principle, the essence of which is that it is unjustifiable to wait for further certainty in the science if the environmental harm involved is irreversible (see below 3.4.2). Moreover, we have seen that scientists predict a highly likely range of *severe* impacts from climate change and a serious possibility of catastrophic harm. The fact that there is a range of possible impacts does not undermine the need for action to mitigate climate change. Consider the following hypothetical scenario. Scientific studies indicate that the release of highly toxic material into a river would very likely cause between 20 and 80 human deaths in a downstream village. The fact that in this scenario there is a *range* of likely harm does not justify breaching the causal chain of ethical responsibility and we would likely all agree that the release of the toxic material should not occur.[12]

The argument presented above for not taking mitigation action is often linked to a claim that one should wait until there is more certainty in the science, and in the meantime increase scientific research funding. This approach is also problematic. If we consider our hypothetical river scenario above, it would not make sense to allow the toxic material to be released in the river pending better scientific data on the likely number of deaths. There is a strong analogy between this scenario and tipping points. In relation to the latter, scientists maintain that if we wait until we have more precise information on the thresholds involved, it will be too late and that therefore early preventive action in reducing GHG emissions is a better option (Allison et al. 2009: 7). It is also important to bear in mind that uncertainty also arises from economic, social and other factors which are built into the models and are important in terms of predicting future emissions (Jamieson 1992).[13]

The argument that climate science is too uncertain to serve as a basis for mitigation is also based on a misconception about scientific method. The assumption is sometimes made that science can deliver absolute (100 per cent) certainty in causal relationships. Scientific method does not operate in this manner. Rather, science deals with working hypotheses that are tested in relation to empirical data with all claims 'potentially open to revision' (Peel 2005: 35). The climate models upon which much climate change science is based have been tested against observations in relation to actual changes in climate change and found to be highly reliable (IPCC 2007b: ch8–9). As outlined above, the key processes in climate change science are known with a high degree of certainty. The evidence is with greater than 90 per cent certainty that human activities have been causing increased global warming from GHG emissions. There is a less than 50 per cent likelihood that the observed increase in global temperature over the last 50 years is occurring owing to factors other than anthropological GHGs (IPCC 2007b: ch9, p665). The idea that science can deliver 100 per cent certainty is based on an 'illusive fiction' (Peel 2005: 35).

Climate change policy involves policy-making in relation to risk of harm, and in this sense is no different from any other environmental policy areas. Science – with its in-built uncertainty – can only present information about likely risks, it cannot prescribe policy responses. Ultimately, policy decisions about what level of climate-change-related risks is acceptable is a question of societal values (Peel 2005: 151).

1.2 GLOBAL MITIGATION REQUIRED: A BUDGET APPROACH

Values are reflected in the defining of a global mitigation objective informed by sound science. The defining of such an objective, which is crucial for delivering intergenerational justice, is vital.

The United Nations Framework Convention on Climate Change (UNFCCC) defines its objective as 'stabilization of GHG concentrations in the atmosphere at a level that would prevent dangerous anthropogenic interference with the climate system' (Article 2). The UNFCCC does not define the point at which 'dangerous anthropocentric interference' might occur, and defining such a threshold is inevitably a policy issue necessarily involving value assumptions (Oppenheimer and Petsonk 2005). Since 2008 a number of international instruments have accepted the requirement of keeping the global mean temperature increase below 2°C (for example Copenhagen Accord).[14] However, scientists have been suggesting that 'the current temperature increase compared to pre-industrial levels (~0.8°C) is already sufficient to increase the likelihood of severe and damaging events occurring' (Min et al. 2011).[15] It is clear that even if emissions were to be kept under 2°C, this would involve considerable harm. For example, with a global temperature rise of 2°C, the area flooded in Bangladesh is projected to increase by 'at least 23–29%' (IPCC 2008: 41).

The IPCC in its Fourth Assessment Report indicated that a 10–40 per cent reduction in global GHG emissions by 2020 below 1990 levels and a 40–95 per cent reduction by 2050 as against 1990 levels was required in order to have about a 50 per cent chance of limiting temperature increases to 2°C above preindustrial levels (IPCC 2007d: 748). More recently scientists have been concluding that a total decarbonisation of the global economy is required to keep global warming below 2°C. For example, Allison et al. (2009: 7) conclude that if climate change is to be limited to a maximum of 2°C above preindustrial levels, global GHG emissions need to 'peak between 2015 and 2020 and then decline rapidly'. Moreover, to stabilise global climate 'a decarbonized global society – with near zero emissions of CO_2 and other long-lived GHGs – needs to be reached well within this century … with average annual per capita emissions … well under one metric tonne CO_2 by 2050', which amounts to a 80–95 per cent reduction in per capita emissions in developed nations compared with emission levels in 2000 (Allison et al. 2009: 7).

The notion of moving to a decarbonised economy entails 'emission budgets'. Put simply, the idea is that GHG emissions are akin to a limited 'resource'. Between now and the time at which the global economy is decarbonised, this 'resource' must be fairly allocated within an overall cap or 'budget'. The link between emissions and carbon concentrations in the atmosphere is analogous to water filling a bathtub: to avoid the bathtub overflowing – exceeding the 'budget' – it is essential to turn off the tap, preferably gradually or this must be done abruptly at the last moment (Saul et al. 2012: 63).

Meinshausen et al. (2009) have analysed GHG emission budgets for the 2000–50 period that would limit warming throughout the twenty-first century to below 2°C. This research builds on the Fourth IPCC Assessment Report and more recent research (Meinshausen et al. 2009: 1158). Their research concluded that for there to be even a 50 per cent chance of staying within a global warming target of 2°C, cumulative emissions from fossil sources and land-use change must be kept within 1000 gigatons (Gt) CO_2 over the period 2000–49. (Meinshausen et al. 2009: 1159). A similar conclusion is reached by Allen et al. (2009a: 1163–6), who conclude that even considering CO_2 alone, cumulative emissions of carbon must not surpass 1 trillion tonnes. They point out that it has taken the last 250 years to burn the first half trillion tonnes of carbon, but on current trends the next half trillion tonnes of carbon will be burned in less than 40 years (Allen et al. 2009b: 57).[16]

The idea of a carbon budget of 1 trillion tonnes is helpful as a 'shorthand' for a climate change mitigation objective (Shue 2011: 302). However, it must be recognised that this particular budget may need to be modified given the increasing likelihood that a 2°C rise in global mean temperature becomes unattainable (UNEP 2011), forcing a shift to, for example, a 3°C target. A further possibility is that changes in the science require a tighter emissions budget, bearing in mind the role that values play in deciding upon a particular emissions target (Shue 2011: 302).[17] In spite of these factors, the underlying logic of an emissions budget remains valid.

The enormity of the challenge of staying within this cumulative budget is evident when one considers that the International Energy Agency (IEA) predicts a global increase in energy demand of one-third to 2035, with fossil fuels remaining dominant in the energy mix, in spite of an increase in renewable energy forms and natural gas (IEA 2012: 1). Global GHG emissions must peak soon; as the risk of exceeding 2°C increases, the later the peak in emissions occurs (Meinshausen et al. 2009: 1158).

How does a budget approach translate to mitigation targets? There is a range of approaches to establishing goals for climate change mitigation, including emissions concentrations in the atmosphere, emissions reduction goals or impact goals. The German Advisory Council on Global Change (WGBU 2009) and Garnaut Review (Garnaut 2008c), building on the work of Meinshausen et al. (2009), have adopted the budget approach described above whereby global mitigation is equated to the optimal depletion of a finite 'resource', in this case the resource being the 'total ecologically tolerable quantity' of GHG emissions possible without triggering dangerous climate change (WGBU 2009: 21). Under this approach a global upper limit for emissions in the atmosphere is specified which is correlated with an overall budget for GHG emissions budgets with national budgets then assigned under the overall global budget. As we will see, this budget approach is particularly well suited to addressing intergenerational justice issues (see below 1.5).

1.3 JUSTICE AND ETHICS ISSUES RAISED BY THE SCIENCE

While the focus of this book is what justice requires of current generations towards future generations in relation to climate change, it is essential to assess this question within the broader questions of ethics and justice raised by the climate science sketched above. By 'ethics' I am referring to questions relating to whether action 'is right or wrong' (Mackie 1977: 9). In this book I use the term 'ethics' interchangeably with the term 'morality'.[18] By 'justice', I am referring to questions that involve the distribution of 'a given measure or advantage (or benefit)' (Page 2006: 51), in other words, who gets how much of the pie, whether the 'pie' be natural resources, welfare or some other benefits. In the climate change context, the matrix of justice can refer to GHG emissions or benefits derived from emissions (discussed in Chapter 3).

The problem of climate change raises a number of justice issues. We have seen that by the middle of this century global GHG emissions will need to be reduced to zero (hereafter 'decarbonised'). If we think of GHG emissions as akin to a 'resource', this gives rise to the following distributional justice issues:

1. *intergenerational*[19] – the level of mitigation burden to be borne by the current generation compared with future generations
2. *intragenerational* – justice within generations, that is, the level of mitigation to be borne by those living today in industrialised countries compared with those in developing countries, and between the poor and the rich within countries
3. *international* – the mitigation to be adopted by states vis-à-vis other states regardless of the distribution within states, including whether this be based on historical emission contributions, emissions since 1990, or other factors (Tremmel 2009: 5).

It should be noted that the level at which justice issues are considered is vital. Thus categories '2. intragenerational' and '3. international' differ in that the latter operates between states whereas the former operates between individuals and societies. The current international climate regime is state-based, with international justice principles that apply between countries embedded in the international legal structure (see 1.9 below). As we shall see, there are tensions between international justice and individual-level justice. For example, there are wealthy elites in developing countries who cause high GHG emissions, and poor people in industrialised countries, for example the poor in the US, who cause negligible emissions. But given the focus in this book on how obligations towards future generations should be reflected in international law, the focus is on categories (1) *intergenerational justice* and (3) *international justice*.

1.4 INTRAGENERATIONAL AND HISTORIC JUSTICE

It is important to understand that addressing the intergenerational justice issues, which are the focus of this book, necessarily involves issues of intragenerational justice, namely the distribution of mitigation burdens amongst people alive today. These distributional issues involve *historic justice* arguments, for example, the argument made by Brazil (and others) that industrialised countries whose emissions since the Industrial Revolution have caused the bulk of total emissions today have a correspondingly larger mitigation burden.[20] Indeed, fossil fuel CO_2 emissions from industrialised countries contributed cumulatively about 79 per cent of the global temperature increase due to long-lived CO_2 emissions that result from fossil-fuel burning as of 1995 (Höhne and Blok 2005: 159). Recall that other GHGs (for example methane, nitrous oxide and halocarbons) also contribute to the overall warming and that their contribution is also significant to current warming. Consequently, a persuasive argument can be made that industrialised countries have overused what can be considered their entitlement to the atmospheric sink (see 3.5.4 below).

It is important to place these historic emissions in context. Some GHG emissions – such as methane – do remain in the atmosphere for a relatively short time. In contrast, CO_2 remains in the atmosphere for hundreds of years, with 20 per cent of CO_2 emissions still in the atmosphere after 650 years (Höhne and Blok 2005: 156). Nearly half of global cumulative GHG emissions have occurred since 1980. Moreover, the burgeoning growth of developing country emissions over the next decades will mean that pre-1990 emissions will represent a rapidly shrinking proportion of global cumulative emissions (Höhne et al. 2011) and thus by 2050 the fossil fuel CO_2 from industrialised countries is estimated to contribute about 55 per cent to the projected global temperature increase under the IPCC SRES A2 scenario. If one considers all GHGs, industrialised countries are projected to contribute 45 per cent and developing countries 42 per cent by 2050 (Höhne and Blok 2005: 159). Under the same model, developing countries will have overtaken developed countries by 2050 in terms of radiative forcing – the impact on increased temperature on the surface of the earth – in relation to all gases, with industrialised countries 34 per cent and developing countries about 48 per cent. However, per capita emissions will remain much higher in industrialised countries. Even as of 1995, industrialised and developing countries already had equal contributions to current radiative forcing that drives the surface temperature changes. The projected relatively high emissions of industrialised countries reflect their earlier emissions and the longevity of CO_2 in the atmosphere.

Even though most cumulative GHG emissions have occurred since 1980, historic emissions are important. For example, if one only counted emissions from 1900, one would ignore roughly 10 per cent of the cumulative effect of GHG emissions (Höhne and Blok 2005: 158). Taking into account historic emissions in mitigation burdens would only be possible if reasonably reliable data was available. The measuring of global historic CO_2 emissions is not perfect but roughly comparable compared with the reliability of current measurement of emissions. However, measuring the contribution of individual countries is more problematic, particularly where those countries have an unusual emissions profile (Höhne et al. 2011: 8). Furthermore, GHG emissions data relating to methane and nitrous oxide have higher levels of uncertainty (Höhne et al. 2011: Annex 3) relative to fossil fuel CO_2 emissions. Considerable uncertainty also remains in relation to measuring accurately historic emissions from forestry and land-use change (Höhne et al. 2011: 6).

The upshot of all of this is that historic GHG emissions should be appropriately taken into account in climate change mitigation and adaptation policies. But the shrinking proportion of global temperature rise attributable to historic emissions of industrialised countries suggests that care should be taken in terms of how this is done. Following Pickering and Barry (2012), I argue that historic emissions should be taken into account in funding mechanisms that assist developing countries in mitigation and adaptation. This is preferable to taking into account historic emissions in designing mitigation targets for individual countries that face the challenges outlined above. Further challenges in the latter include the difficulty of taking into account the fact that developing countries in the more recent period have been able to benefit from improved, more efficient technologies, which industrialised countries did not have access to during their development, although barriers to technology transfer to developing countries cannot be overlooked (Höhne and Blok 2005: 164).

Some experts argue that historic emissions of industrialised countries provide a strong rationale for allowing an overshoot in emissions by developing countries even beyond equal per capita shares (Garnaut 2008c: ch9, 9.5). However, allowing an overshoot can arguably jeopardise the possibility of reducing global GHG emissions rapidly enough in order to protect the interests of future generations (Macintosh 2010).[21] Very quickly we can see that intragenerational justice issues impact on intergenerational justice and the two cannot be considered in isolation.

1.5 INTERGENERATIONAL JUSTICE

Intergenerational issues arise because a failure to make deep GHG emission cuts now will mean deeper and more economically disruptive reductions being required in the future, for example in 2030 when another generation is in positions of political power. See Figure 1.1.

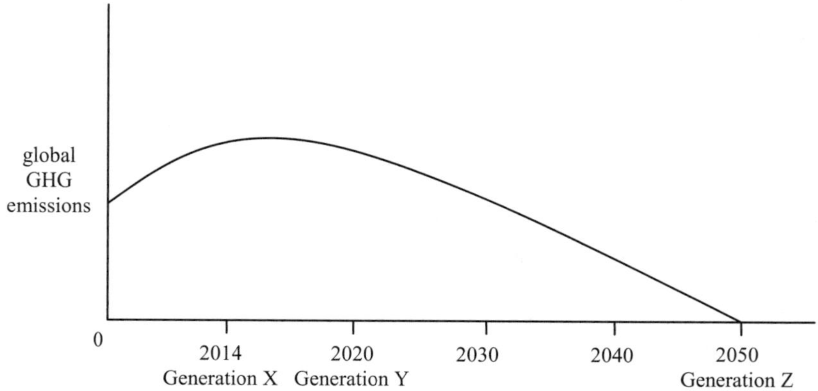

Figure 1.1

Another way of describing this is that we must devise a fairness or justice principle for dividing up a limited 'resource' consisting of GHG emissions (Shue 2011) – the area under the curve in Figure 1.1 – from now until roughly 2050, when the global economy must be decarbonised. Whatever distributional principle we adopt must operate within a context of interdependent generations, where for example action by Generation X (in 2014) in sharply reducing GHG emissions mostly benefits Generation Z (in 2050). The flipside of this is that where Generation X *fails* to act in reducing sharply GHG emissions, Generation Z will face higher costs in reducing emissions and greater risk of significant climate-change-related harm. In this model, the appearance of discrete generations and their labelling for convenience Generations 'X' and 'Z', is a simplification given the nature of overlapping generations (see below 1.5.1).

It is evident from the above that intergenerational justice issues arise in relation to climate change adaptation as well as mitigation. A failure by the current generation to strongly mitigate climate change will result in more harmful climate change impacts than would otherwise occur and increased adaptation costs to future generations.

Tremmel (2009: 19–20) points out that the term 'generation' is ambiguous and can refer to (1) people sharing the same family lineage, (2) a group of people with shared beliefs – societal generations, (3) a particular age group, that is, the young or middle-aged or elderly in a society alive at the same time or (4) everyone alive today. Tremmel distinguishes between the third and fourth categories, which he describes as 'chronological-temporal' and 'chronological-intertemporal', which are often blurred in other theories. In his theory he uses the term 'temporal justice between generations' for 'justice between young, middle-aged and older people alive today', whereas 'intertemporal generational justice' is defined as 'justice between people who lived in the past, people alive today and people who will live in the future' (Tremmel 2009: 22). In this book – unless stated otherwise – 'intergenerational justice' is used with the same meaning that Tremmel assigns to intertemporal generational justice. In this book 'future generations' will be defined (following Tremmel 2009: 24) as referring to generations where 'none of its members is alive at the time the reference is made' with all of its members born after the reference was made.

1.5.1 Cascading Generations and Catastrophic Impacts

Generations do not constitute a discrete group with one generation following the other in a neat chronology. Rather, generations are overlapping (Barry 1977: 271). This particular cascading quality of generations does not prevent one from making an assessment in very general terms about the respective interests of young people, people in positions of power or people to be born in the future at a particular point in time.

Prior to the climate change issue becoming so acute, philosophers tended to argue that current generations had a particular responsibility to their immediate successor generation; however, it was considered difficult to extend such obligations well into the future beyond this up to, for example, 100 years. Climate change science has forced a change in these arguments for a number of reasons.[22] Firstly, GHG emissions have a long lifespan, remaining in the atmosphere for long periods. A consequence of this is that a failure to sharply reduce GHG emissions now will have an impact on not just the next generation but succeeding generations after that. In addition, as we have seen, scientists are concerned that we risk reaching tipping points whereby catastrophic and irreversible changes in the climate system occur. Given the significant risk of this occurring, the current generation bears a particular responsibility.

1.6 SUBSTANTIVE AND PROCEDURAL JUSTICE

So far the discussion has been in terms of substantive justice. However, procedural justice is also important. The essence of procedural justice is strongly linked to a democratic principle that '[p]roblems that affect some group of persons ought to be decided by those persons, either directly or through their representatives' (Vanderheiden 2008: 90). Thus in the making of international treaties there must be meaningful participation of those affected by the treaty in the processes involved to ensure that their interests are taken into account. An important rationale for this principle of participation is that fair substantive policies are more likely to occur if the policy-making process itself is fair (Vanderheiden 2008: 62). This participatory principle is reflected in Principle 10 of the 1992 Rio Declaration, which states that '[e]nvironmental issues are best handled with the participation of all concerned citizens, at the relevant level'. The Aarhus Convention embodies this notion in a particularly strong form, linking an obligation on parties to guarantee 'public participation in decision-making' as a prerequisite for 'the protection of the right of every person of present and future generations to live in an environment adequate to his or her health and wellbeing'.[23]

Procedural justice is particularly important in international law, as international treaties that are perceived to be fair in terms of the processes by which they are made are more likely to be complied with (Alford and Tierney 2012: 17). Given international law's lack of centralised enforcement mechanisms, procedural justice can help ensure legitimacy with better chances of compliance (Franck 1995: 26).

This book seeks to identify criteria for attaining substantive justice for future generations in relation to climate change. But parallel to this, I argue that procedural justice should be strived for, as it is more likely to deliver substantive justice, measured through fulfilment of the proposed criteria. Given that future generations cannot directly take part in international law-making affecting their interests, 'rough' procedural justice can only ever be attained, based for example on indirect representation of future generations' interests by proxies. Such mechanisms are inherently difficult given the likelihood of current pressing concerns swamping the interests of future generations. Moreover, there is tension between substantive and procedural justice in the climate change context. More inclusive treaty-making processes may be procedurally fair but come at a price in terms of substantive justice, as the greater the number of parties involved in negotiations, the more difficult it is likely to be to reach agreement (Eckersley 2012: 25). This is problematic as reaching agreement quickly on mitigation action is crucial for future generations.

Chapter 7 briefly explores how intergenerational procedural justice can be incorporated in international law. This includes a discussion of potential international mechanisms with an explicit mandate to represent future generations' interests. Such mechanisms exist in some countries at the national level in the form of, for example, sustainable development commissions (Weston and Bach 2008). Indeed, during the negotiations leading to the Rio+20 Conference during 2012, a 'high level representative for sustainable development and future generations' was considered and strongly supported by the EU and many other Member States.[24] However, it did not make its way into the outcome of that conference owing to concerns by some countries that such a mechanism would be used to monitor governments and that their national sovereignty was under threat.[25]

While procedural justice is important – and procedural mechanisms are worth exploring at the international level – a full evaluation of such proposals requires the input of social scientists and is beyond the scope of the normative focus of this book. This book, however, does provide a normative framework which could be valuable in considering procedural options being proposed.

1.7 WHY DOES INTERGENERATIONAL JUSTICE MATTER FOR POLICY-MAKERS?

As Garnaut points out, '[a]ction to mitigate climate change is premised on an obligation being owed to unborn generations as well as younger generations currently alive' (Garnaut 2008a: 5). So clarity on the nature and scope of such an obligation is an important precondition for effective policy-making in this area. Moreover, intergenerational justice is built into the current international legal framework and negotiating mandate for further developing this framework (1.10 and Chapter 4 below). In this sense, justice concerns are required to be considered by policy-makers (Soltau 2009: 3).

The focus in this book is on climate change mitigation, in other words action to reduce GHG emissions. Issues of how human societies adapt to climate change – so-called issues of adaptation – are also important but outside the scope of this book. The reason for limiting the book to mitigation issues is that mitigation is the 'first best' response to climate change. Delayed mitigation action entails passing on to the next generation greater costs in climate change adaptation. So intergenerational justice is involved in both climate change mitigation and climate change adaptation policy-making.

1.8 ECONOMICS AND DISCOUNTING THE FUTURE

Unmitigated climate change will involve substantial economic damage for future generations. The Stern Review concluded that unmitigated climate change would involve a decrease in the order of global GDP of between 5 per cent and 20 per cent (Stern 2007).[26] A cost–benefit approach would therefore suggest that it is worthwhile to take strong climate change mitigation action now.

However, any economic cost–benefit analysis in relation to climate change mitigation must make an assumption about the value of damages in the future in dollar values today. Economists describe the method of converting future damages into contemporary dollar values as discounting (Splash 2002: 203). A key component of discounting is the notion of 'pure time preference', which refers to the idea that generally people prefer to have goods now rather than in the future and takes into account the risk that a person may not exist at some point in the future (Cline 2004: 5).

Many economists have used high discount rates to justify postponing strong mitigation action to address climate change (for example Nordhaus 2008).[27] To illustrate, the cost of adapting to a 2°C warmer world by 2050 is estimated to be in the range of 70–100 billion USD per year by 2050 (World Bank 2010). Applying a discount rate of 6 per cent would value this today at only 17 400 USD. Thus the value of future damage looks miniscule in today's values and pales in comparison to mitigation costs. Stern reflects the approach of ascribing equal value to all persons and, since his influential report, there has been a move away from high discount rates towards low discount rates in relation to climate change damages.[28]

High discount rates conflict with a *core human rights principle* according to which all persons born now and in the future have an equal right to life, health and subsistence (elaborated at 3.5.1 below). This principle entails that persons born now and in the future are of equal value. This book supports an interest-based theory of rights (Chapter 2), according to which there is no space for pure time preference as '[T]ime is not a morally relevant consideration' (Caney 2009: 169). The notion of 'pure time preference' entails in effect discriminating against people on the basis of when they happen to be born (Caney 2009: 169). Such discrimination violates a principle of 'impartiality' according to which political decisions should not penalise people on the basis of properties 'that lack any fundamental moral relevance' (Caney 2009: 168).

There are three additional problems with applying high discount rates to the cost–benefit analysis of mitigation measures. Firstly, the fact that individuals in their decisions relating to consumption give priority to immediate needs rather than long-term interests does not mean that public policy *should* follow such an approach (Splash 2002: 211). What individuals do in practice does not necessarily correspond to what they *ought* to do. Secondly, high discount rates, as we have seen, convert into weak climate change mitigation action, which leaves future generations threatened with a significant risk of catastrophic harm (Weitzman 2007: 707). So high discount rates entail ineffective policy prescriptions. Linked to this point is the assumption of continued economic growth, which may be questionable in the long term (Meadows et al. 2004) or even the short term (Carmody 2012: 69).

A related argument made by Lomborg (2007: ch24) and others is that expenditure on strong climate change mitigation now is misplaced because it involves ignoring the contemporary global poor and preferences future generations vis-à-vis current generations, which is unfair, as future generations will be wealthier than contemporaries.

However, the opportunity cost argument of Lomborg and others rests on an assumption that destruction of the global ecological system can be substituted by an increase in economic wealth. Put bluntly, if we pass onto the next generation a seriously impaired climate system, increased levels of wealth, education and technology will not be a satisfactory compensation for this radical change to a key earth system (Neumayer 1999: 34). This assumption of 'substitutability' is dubious as it ignores human beings' dependence on the global ecological system (Neumayer

1999: 34), and in any event breaks down when the existential threat of climate change to the continued functioning of human civilisation is appreciated.

1.9 WHY FOCUS ON INTERNATIONAL LAW?

This leads to the question of why international law is important for addressing intergenerational justice aspects of climate change. Put differently, is a focus on international law in this book justified? An international treaty regime for addressing climate change is essential for a number of reasons. Firstly, without such a treaty regime, states which take action will bear costs not borne by their trading partners, which can lead states to avoid taking any action at all (Miller 2008: 121). Secondly, an international regime is essential in mobilising technology and funds to assist in both mitigation and adaptation efforts (Philibert 2004). To be most effective, technology transfer needs to be linked to binding targets under a climate change regime (Lawrence 2007: 195).

Reaching some broad agreement internationally on what constitutes both intergenerational justice and intragenerational justice is arguably a precondition for an effective international treaty addressing the climate change problem. Given the difficulty in enforcement mechanisms for international treaties, fairness is particularly important, as an agreement which is considered fair is more likely to be complied with (Miller 2008: 123). The US (and the previous Australian government under John Howard) argued that the Kyoto Protocol was *unjust* as it did not place any restrictions on emissions from developing countries. Even if such arguments were motivated by economic self-interest, there is no doubt that such arguments influenced the policy direction of these countries and viability of the treaty regime. Thus fairness can influence whether a treaty regime comes into being in the first place (Page 2006: 113).

The 'realist' school of international relations argues that concepts of justice are just window dressing, with power politics between nations pursuing their own sovereign national interests being determinative. However, this version of international relations glosses over the complex interaction between hard economic interest and justice principles involved in international negotiations (Scholtz 2009: 11–13). Thus a recent survey of responses of people involved in international negotiations 'concluded that equity was considered important' (Lange et al. 2007: 549–62).[29] Moreover, to the extent that states act for selfish nationalist reasons, this does not mean that they *should* act in this way (Vanderheiden 2008: 94). Vanderheiden points out that 'realists' commit a so-called 'naturalistic fallacy' by deriving an *ought* from an *is;* 'even if all persons really are egoists at heart, this does not make it right or proper for nations to disregard ideals such as justice' (Vanderheiden 2008: 95).

1.10 THE CURRENT UN CLIMATE REGIME AND NEGOTIATIONS

We have seen that climate change science requires a virtual decarbonisation of the global economy by 2050, with GHG emissions reduced 10–40 per cent by 2020. How does the current international legal regime measure up to this imperative? The 1992 United Nations Framework Convention on Climate Change (UNFCCC) established a broad framework for cooperation and an overall goal of avoiding dangerous anthropogenic climate change (Article 2), but did not establish binding mitigation targets. The 1997 Kyoto Protocol to the UNFCCC did establish binding mitigation targets but only on industrialised countries and economies in transition (Eastern European countries). However, even if fully implemented, the Kyoto targets will only deliver a 5 per cent reduction in global emissions for industrialised countries and economies in transition (Eastern European countries) for the period 2008–12 (Sands and Peel 2012: 286).

In December 2009, at the Copenhagen conference of parties (COP) to the UNFCCC, governments agreed on a global objective of keeping global warming below 2°C (Copenhagen Accord para 2). This was accompanied by voluntary mitigation targets announced by many countries at this conference or in the period following. However, these efforts fall well short of what is required to reach a 2°C target (UNEP 2011). Moreover, to date, effective mitigation targets as part of the global climate regime remain elusive. The 17th COP to the Kyoto Protocol, held in Durban South Africa in 2011, made an in-principle decision to extend the Kyoto Protocol for another five to eight years (2013–2017/20). Parallel to this, the Durban conference agreed to launch under the so-called 'Durban platform' a negotiation process to develop a 'protocol, another legal instrument, or agreed outcome with legal force applicable to all parties' addressing the post-2020 period.[30] But Durban failed to deliver an agreement on when global emissions should peak or any long-range global emissions reduction target. Moreover, the negotiating timetable – with a global agreement to be negotiated by 2015 and enter into force in 2020 – flies in the face of the urgency required by climate change science to deliver justice for future generations (Bodansky 2012a).

At the meeting of parties to the Kyoto Protocol held in Doha in 2012, an amendment to the Kyoto Protocol was agreed enabling a second commitment period for the Protocol from 2013 to 2020.[31] However, upon entry into force, the binding emission targets contained in the amendment will only apply to the EU, Australia, Switzerland and a few industrialised countries, totalling only 15 per cent of global GHG emissions.[32] The US, Japan, Canada and Russia are not included.

It is important to note that while concrete mitigation action has been weak to date, the principle of intergenerational justice is reflected in Article 3(1) of the UNFCCC, which provides:

> The Parties should protect the climate system for the benefit of present and future generations of human kind, on the basis of equity and in accordance with their common but differentiated responsibilities and respective capabilities. Accordingly, the developed countries Parties should take the lead in combatting climate change and the adverse effects thereof.

The current UNFCCC negotiations are based on the Durban platform but also the Bali action plan adopted in 2007.[33] The latter includes the intergenerational principle described above as part of the principles in Article 3 of UNFCCC which comprise the relevant mandate. The implementation of these principles will continue to play a role in the negotiation of a post-2012 climate change regime, influencing in various ways the shape of those rules (see Chapter 4).

1.11 CONCLUSION

This chapter has had three broad themes: firstly, the intergenerational justice issues arising from climate change science; secondly, the intergenerational justice issues arising from 'discounting' by economists and related assumptions of increased wealth; and thirdly, the importance of international law as a vehicle for addressing intergenerational justice in the climate change context.

Intergenerational justice issues arise from the science of climate change owing to the persistence of CO_2 in the atmosphere over long periods of time. Climate change science tells us that there is a high likelihood of future generations suffering increased mortality rates owing to climate-change-related impacts, some of which are irreversible (for example ocean acidification and species extinctions). Intergenerational justice issues arise as a failure to mitigate climate change by the current generation entails very likely significant harm to the interests of future generations. It also entails higher costs for future generations in terms of both mitigation and adaptation. Moreover, a failure by the current generation to take strong mitigation action entails a significant risk of catastrophic damage for future generations.

There is overwhelming support amongst atmospheric scientists that human beings are contributing to climate change by GHG emissions, while rates and regional distribution of impacts remain uncertain. But this uncertainty cannot provide a valid justification for delaying mitigation action for two reasons. Firstly, it misunderstands scientific method by incorrectly assuming that science can deliver absolute (100 per cent) certainty. Secondly, mitigation action is justified even if there is a range of harmful outcomes: precision in the level of harm is not a reasonable precondition for taking action when all the risk scenarios involve significant harm. The nature of the problem suggests that a precautionary approach is required: otherwise, action can end up being delayed until it is too late to be effective.

In response to these threats, scientists have concluded that the decarbonisation of the global economy is required by 2050 to have a better than even chance of staying below a 2°C global temperature rise – a widely accepted threshold for avoiding the worst of these impacts. A useful shorthand way to think of the mitigation challenge is to frame the issue as how to distribute fairly a cumulative emissions allocation of 1 trillion tonnes of CO_2 (Shue 2011). This is an enormous challenge given global energy demand is currently continuing on an upward trajectory.

While the focus of this book is on intergenerational justice and international justice (between states), this cannot be considered in isolation from intragenerational justice. Thus arguments relating to whether historic emissions should be taken into account relate to both the role of industrialised countries' historic emissions in causing the climate change problem and the enormous disparities in current levels of wealth between developing and developed countries. Developing countries are likely to have an ever-increasing share of emissions, while per capita emissions of developing countries are likely to remain higher than industrialised countries for some time to come.

We will see that one of the most challenging issues in developing justifiable justice-based principles is how to balance the pressing needs of current generations – particularly the poor – with the needs of future generations. Indeed, economists have used high discount rates to justify delay in strong climate change mitigation action, often combining this with an argument that future generations will be wealthier than current generations, thus more capable of addressing climate change (for example Lomborg 2007). But these approaches make assumptions which conflict with notions of persons being born now and in the future being of equal worth. Furthermore, these approaches assume that permanent damage to global ecosystems can be replaced by increased economic wealth.

This raises the issue of whether procedural mechanisms can be devised to give future generations a greater voice in the law-making process in order to avoid permanent damage to global ecosystems upon which human beings depend. We observed that recent proposals to include such an international mechanism in the Rio+20 Conference ran into sovereignty concerns.

Indeed, state sovereignty remains at the heart of international law, leading some to suggest that international law is structurally incapable of addressing the intergenerational justice issues raised by climate change. But only a binding international treaty-level instrument is capable of addressing trade competitiveness concerns and delivering the necessary stability of commitments into the future, essential for intergenerational justice. The focus of this book on international law is therefore justified.

However, the current climate treaty regime only weakly reflects the imperatives of climate science and the need for action to ensure intergenerational justice. The Durban mandate is inadequate in its time frame, with the new global climate treaty to enter into force in 2020. The Kyoto Protocol has limited coverage (only 15 per cent of global emissions) and inadequate mitigation commitments.

Developing a global climate treaty under the Durban mandate depends on at least some broad agreement on the requirements of justice, both *intra-* and *inter-*generational. In the next chapter we turn to the latter.

NOTES

1. In this book the terms 'intergenerational justice' and 'intergenerational equity' will be given the same meaning unless it is clear from the context that the international law principle of 'intergenerational equity' is being referred to. The latter is addressed in Chapter 4.3.1 below.
2. See ABC News (2008), 'Government encouraged to welcome "climate change refugees"', ABC News, Mon 14 July, available at http://www.abc.net.au/news/stories/2008/07/14/2302737.htm (accessed 19 December 2008). Inhabitants of the Carteret Islands in Papua New Guinea also face the prospect of relocation.
3. See Stern (2007), Garnaut (2008c: ch9).
4. Neither the United Nations Framework Convention on Climate Change nor the Kyoto Protocol defines the terms 'developed' or 'developing' countries. It is clear, however, that 'developed countries' are equated with UNFCCC Annex 1 parties and 'developing countries' with non-Annex 1 parties. The terminology used in this book follows this understanding while recognising this is problematic as some 'developing countries' so defined include South Korea, Mexico and Chile, who are members of

the OECD (D. Bodansky, 'W[h]ither the Kyoto protocol, Durban and beyond' August 2011 Harvard Project on Climate Agreements http://belfercenter.ksg.harvard.edu/files/Bodansky_Viewpoint-Final.pdf. Accessed 13 September 2013).

On occasion I use the term 'industrialised' synonymously with 'developed'. See 7.4 below, which discusses how the classification of countries under the current UNFCCC may need to be reformed.

5. But developing countries have nevertheless also emphasised obligations towards future generations in their rhetoric; see Chapter 6 below.
6. Scholtz (2009: 167) uses the word 'pragmatism' but I prefer the term 'feasibility' as it captures the notion of whether a climate regime is feasible in addressing the environmental issue at stake.
7. There is considerable literature on sustainable development in international law (see 4.3.1 below), but very little material on intergenerational equity as a freestanding concept in international law. The notable exception is the path-finding book by Edith Brown Weiss, *In Fairness to Future Generations: International Law: Common Patrimony, and Intergenerational Equity* (The United Nations University). See also E. Agius and S. Busuttil (eds), *Future Generations and International Law* (Earthscan 1998).
8. The fifth IPCC Assessment Report Working Group 1 The Physical Science Basis is due to be released in September 2013, Working Group 2 on Impacts, Adaptation and Vulnerability, March 2014 and Working Group 3 on Mitigation, April 2014 (www.ipcc.ch accessed 27 March 2013).
9. Other surveys include Oreskes, 'Beyond the Ivory Tower' (2004), and Doran and Kendell-Zimmerman, 'Examining the Scientific Consensus on Climate Change' (2009).
10. In Allison et al. (2009: 34).
11. Some have suggested that the sheer scale of human-generated changes to the biosphere means that a new geological era is beginning and that 'the relatively stable geological era which prevailed since the previous ice age – the Holocene, no longer applied'. The term 'Anthropocene' has been used to describe this transformation (Simon Dalby, 'Biopolitics and Climate Security in the Anthropocene', 49 (2013) *Geoforum* 184). See also references in Louis Koetze, 'Reimagining Global Environmental Law and Governance in the Anthropocene' 27 September 2012, Scientific Contributions, Series H: Inaugural Address: 252 http://dspace.nwu.ac.za/handle/10394/8575 accessed 2 September 2013, 3.
12. This scenario assumes that the activity producing the toxic material can be carried out in a manner that does not produce the toxic material, that is, there are viable less harmful options. How this assumption plays out in relation to climate change is discussed below at 3.5.3.
13. In: Gardiner et al. (2010: 81).
14. Decision 2/CP15, Report of the Conference of the Parties on its Fifteenth Session, Copenhagen, 7–19 December 2009, FCCC/CP/2009/11/Add.1 (30 March 2010) (Copenhagen Accord) para 2.
15. Referred to in Garnaut (2011: 38).
16. See http://trillionthton.org/. Accessed 3 October 2012.
17. Thus the WGBU proposes a global budget of 750 gigatonnes of CO_2 from fossil fuel emissions for the period 2010–2050 (WGBU 2009: 2).
18. Grayling points out that 'ethics' is a broader concept than 'morality' involving how a person lives their life, whereas morality is confined chiefly to interpersonal relationships. See A.C. Grayling, *The Choice of Hercules, Pleasure, Duty and the Good Life in the 21st Century* (Phoenix 2007: 71).
19. See 1.5 below for a definition of future 'generations'.

20. For a recent version of the Brazilian proposal, see 'Ideas and proposals on the elements contained in paragraph 1 of the Bali action plans' submissions from parties' Part I (19 May 2009) FCCC/AWGLCA/2009/MISC.4 (Part I) 54–5.
21. Macintosh (2010) suggests that if developed countries adopt a combined target of less than 20 per cent reduction by 2020 against 1990 levels, this would mean that global carbon emissions would have to be reduced post-2030 by between 5 per cent and 10 per cent per year to keep global warming below 2°C. He suggests that the chances of staying below the 2°C target are very low without stronger mitigation targets being adopted. See also UNEP (2011).
22. This reflects the fact that since the nuclear age human beings have for the first time in history the capacity to irreversibly damage the future of the planet by use of technologies (Tremmel 2009: 1).
23. Principle 1 Aarhus Convention.
24. See proposal contained in UN document: 'The Future we Want', United Nations, Rio+20, United Nations conference on sustainable development, 10 January, 2012, page 10, para 57. This is discussed below at 7.6.1.
25. Anonymous source on file with the author.
26. See discussion of Stern's discount rate in Weitzman (2007: 708).
27. Skott and Davis (2012: 1) state that 'Nordhaus (2008) and most other studies … use a "descriptive" approach in which the welfare function of the representative agent has to be calculated to fit empirical observations.' This is contrasted with Stern's 'proscriptive' approach. The proscriptive/descriptive terminology is used by Arrow et al. (IPCC 1996: chapter 4).
28. Nicholas Stern applied a pure time preference of 0.1 per cent to take into account the possibility that the human race is extinguished, for example by a nuclear conflagration which equates to a 90 per cent probability of humanity surviving a 100-year period. He applied an elasticity of marginal consumption rate of 1 and a per capita growth rate of consumption of 1.3 per cent. These figures combined to give Stern a total discount rate of 1.4 per cent (Stern 2007: 663). See discussion of Stern's discount rate in Weitzman (2007: 708).
29. In Kverndokk and Rose (2008: 38).
30. Decision 1/CP.17 para 2, 'Report of the Conference of the Parties on its 17th session, held in Durban from 28 November to 11 December 2011' UN Doc. FCCC/CP/2011/9/Add.1, 15 March 2012.
31. FCCC/KP/CMP/2012L.9 (8 December 2012).
32. R. Harrabin (8 December 2012) 'UN climate talks extended Kyoto Protocol, promise compensation', BBC News, www.bbc.co.uk/news/science-environment/20653018 accessed 14 February 2013.
33. Bali Action Plan, Decision 1/CP.13 adopted by the UNFCCC 13th Conference of Parties, 3–15 December 2007, FCCC/CP/2007/6/ Add. 1 (14 March 2008) http://unfccc.int/resource/docs/2007/cop13/eng/06a01.pdf accessed 18 September 2013.

PART I

Theory

2. The basis of an obligation towards future generations in justice and ethics in the context of climate change[1]

> The climate change problem is at its heart an ethical problem … The main impacts are longer-term ones. The most important potential impacts are very long-term ones. And you have to value the welfare of future generations to want to do anything about this problem.[2]

2.1 INTRODUCTION: KEY ASSUMPTIONS

We saw in Chapter 1 that climate change raises challenging issues of intergenerational justice and ethics. The central question for this chapter is to examine whether there is an ethical obligation of contemporaries towards future generations in relation to climate change mitigation. The approach in this chapter is to draw on a number of theories of justice and ethics. One of the key arguments is that future generations possess human rights, in the sense of moral rights, with an obligation on current generations to protect these rights. My argument draws on theories of justice, including the 'capacity approach' to justice of Sen (2009) and Nussbaum (2006), which has received much attention in recent decades. Also addressed are justice theories of reciprocity, communitarianism and cosmopolitanism and theories based on impartiality such as that of John Rawls.

Ultimately, my argument rests on a number of key assumptions in these various theories of ethics and justice, for example, the notion that persons are of equal value regardless of when and where they are born. My analysis makes transparent these key assumptions in the various theories relied upon. Deeper justification of these assumptions is beyond the scope of this book.

A key argument in this chapter is that an obligation towards future generations rests on future generations possessing core human rights which are threatened by climate change. My approach follows Caney in arguing that in relation to at least core human rights – the rights to life, subsistence and health – 'each and every individual' as human beings is entitled to minimum standards of treatment regardless of where and when they live (Caney 2010a [2010]: 165). Thus one *ought* to respect these core human rights in order to ensure a minimum level of human dignity.[3] (See 2.3.1 below as to why these rights and not others are selected.) These rights comprise moral rights flowing from the value of 'recognising respect for the dignity of every human being'.[4] 'Respect for the inherent dignity of human persons' is reflected in human rights instruments such as the Universal Declaration on Human Rights (UDHR) and various human rights treaties. Thus, the Preamble to the Universal Declaration of Human Rights proclaims that: 'Whereas recognition of the inherent dignity and of the equal and inalienable rights of all members of the human family is the foundation of freedom, justice and peace in the world'.[5] At the heart of the concept of 'respect for human dignity' is the notion that 'each human being possesses an intrinsic worth that should be respected' (McCrudden 2008: 723).[6]

My premise of 'respect for human dignity' linked to core human rights rests on these notions comprising widely shared values in today's world. Donnelly points out that today 'human rights are backed by the predominant political, economic, and cultural powers that have become ideologically hegemonic in international society' (Donnelly 2007: 282). Put negatively, no states in the world proclaim the right to arbitrarily deprive their citizens of physical integrity (Tomuschat 2008: 95).[7] While human rights are asserted as being 'universal', Donnelly shows that human rights are better considered as being '*relatively* universal' by carefully analysing the different senses of 'universality'. He points out that human rights are 'conceptually universal' in the sense of human rights being regarded as 'inherent' in all human beings, which does little more than make universality part of the definition of human rights (Donnelly 2008: 282, Renteln 1990: 49).[8]

Human rights do, however, also have 'international legal universality' in that virtually all states accept the authority of the Universal Declaration on Human Rights and the six key international human rights instruments have been ratified by most states in the international community (Donnelly 2007: 288). This tells us that political elites around the world accept human rights as authoritative but not necessarily that human rights are regarded as authoritative at the societal level.[9] The latter is notoriously difficult to assess (Tomuschat 2008: 89). While non-Western countries had a significant impact in the negotiation of the UDHR,[10] the negotiation of the major international human rights instruments was the outcome of predominantly Western countries' influences (Bell 2000: 64).

Some have argued that human rights are universal in an anthropological or historical sense in that non-Western countries have their own religious and philosophical traditions which reflect values which are common to core human rights (Ishay 2004: 15–61). For example, Confucianism contains the notion of 'ren' involving the idea that one has a duty to avoid human suffering even in relation to persons one does not know (Bell 2000: 50). Sim (2004: 341, 348) shows that while Confucius did not use the language of rights, theories of rights which emphasise that rights are embedded in social relationships are consistent with Confucian ideas.[11] Similarly, Sen has traced in the Indian philosophical tradition values which closely mirror elements of international human rights (Sen 2004: 352).

However, these arguments tend to conflate values such as justice, fairness or humanity with the very specific notion of human rights entailing a realisation of these values (Donnelly 2007: 284).[12] This realisation is to occur through institutions whereby all persons are considered to be of equal worth and having rights against the state. This notion of human rights happened to develop in the West in the Enlightenment but has spread globally and been infused in different cultures, providing a basis for responding to oppression linked to modernisation.[13]

In recent decades, human rights have become universal in the sense that a growing number of worldviews, such as Islam, Kantianism and Confucianism, have come to support human rights as part of a political conception of justice (Donnelly 2013: 57–60). While so-called 'Asian values' have been argued as conflicting with 'Western' human rights, in the recent period values found in a number of Asian countries including Japan, Taiwan and South Korea have been supportive of human rights (Donnelly 2007: 290). China has, since the establishment of the UN Human Rights Council in 2006, formally accepted the universality of human rights while emphasising priority to 'survival' rights of subsistence (socio-economic rights) over civil and political rights (Sceats and Breslin 2012: 8).[14] China sees human rights in an 'aspirational' rather than legal sense, with collective interests emphasised over those of the individual.[15]

It should be noted that the core human rights of life, health and subsistence are not central to the 'Asian values' debate, which has largely involved human rights associated with Western liberalism, including rights to freedom of expression, democracy and freedom of religion (Langlois 2001: 12).[16] These human rights are not relevant in terms of the core human rights threatened by climate change.

Finally, it is important to clarify that my pragmatic approach of identifying as a key premise 'respect for human dignity', and core human rights flowing from this value, is compatible with a range of views on the ontological foundation of human rights, whether based on theology, natural law or utility.[17] Indeed, a range of moral theories relating to human rights is arguably consistent with an approach based on human dignity (Moellendorf 2009a: 11). This pragmatic dimension to my approach is important, as identifying principles which reflect shared values is vital in the task of building an effective climate change regime.[18]

This chapter is structured as follows. The essential argument for an ethical obligation on current generations towards future generations in the climate change context is presented: on the basis of a harm avoidance principle (2.2) and an argument that future generations have human rights (2.3). The analysis then turns to various influential theories of justice. Thus the capacity approach of Sen and Nussbaum (2.4) and a communitarian argument based on the notion of a community extending into the future provide a basis for an obligation towards future generations (2.7). The latter theory is combined with cosmopolitanism justice – the notion that obligations are owed to individuals beyond national borders (2.8). Within the sections outlined above, I deal with various objections. For example, an obligation to avoid harm is argued by some as involving an unsound notion of collective responsibility (2.2.2). Similarly, an argument based on human rights is argued by some – in my view incorrectly – to be invalid on the basis that future unborn persons cannot possess rights (2.3.3(a)). Other sections of the chapter deal with arguments for intergenerational justice which in my view are not convincing. These include arguments based on reciprocity, which are problematic owing to the one-way relationship between contemporary and future generations (2.5), and Rawlsian theories, which pose difficulties owing to their in-built notion of progress (2.6). Finally, I address stewardship (2.9) and sustainability (2.10) approaches, which I argue are good vehicles for implementing intergenerational justice but cannot in themselves establish obligations towards future generations.

Having established in this chapter an overarching obligation on contemporary generations towards future generations in relation to climate change mitigation, Chapter 3 seeks to sketch out its content in terms of principles which address in further detail the intergenerational distributional justice aspects.

In the analysis which follows, I test various justifications for an obligation towards future generations in relation to action to address climate change. Testing of these theories in relation to other environmental issues, such as biological diversity, is an important task beyond the scope of this book.

It is important to note that my arguments for an obligation towards future generations in the climate change context rest on anthropocentric assumptions. In making this argument I do not take a position in denying that a similar argument can be made resting on deep ecology assumptions (2.11 below). Rather, my argument is that an anthropocentric approach resting on the dependence of human beings on the global ecological system – including a stable climate – is sufficient to ground an obligation towards future generations, which, if implemented, would ensure that the global climatic system is kept intact for the benefit of future generations and the global ecological system as a whole. Moreover, an anthropocentric approach is more likely to find resonance across a range of cultural and religious traditions, thus facilitating a global consensus to address climate change (2.9 below).

2.2 OBLIGATION TO AVOID HARM

Intergenerational justice can be based on the extension into the future of an obligation to refrain from causing future harm to other human beings (Vanderheiden 2008: 137). My formulation of this 'harm avoidance principle' is as follows:

> The current generation, particularly those in positions of power, have an ethical obligation to refrain from action which has a high probability of causing serious harm to the basic interests and core human rights of (1) the current generation and (2) future generations.[19]

'Harm avoidance' as a key ethical principle has a long history. John Stuart Mill argued that the prevention of harm to others governed the relationship between the individual and society and was the only basis upon which 'power can be rightfully exercised over any member of a civilized community, against his will'.[20] A similar concept underlies the law of torts and criminal law, although they concern remedying harm with hopefully some deterrent effect: the harm avoidance principle is more directly aimed at preventing harm. In order to justify my formulation of the principle, it is interesting to posit the contrary principle, which would be: 'the current generation has a right to take action which has a high probability of causing serious harm to those born in the future'. This would be problematic as this would arguably justify actions which could threaten the ongoing survival of significant numbers of human beings. Moreover, the harm avoidance principle is consistent with the instinct and/or responsibility to nurture children and parents' efforts in maximising the likelihood of a successful future for their children.[21]

A few points need to be made to clarify the 'harm avoidance principle' set out above. Firstly, the reference to 'those in positions of power' refers to political power. My focus will be on ethical responsibilities of governments rather than companies or individuals (see below 2.3.3(d)). I will be interpreting the principle as extending to the failure of governments to regulate GHG emissions, even if most emissions occur through private rather than governmental activities. In this sense, the failure to take 'action' includes the failure to regulate. The focus on governments is justified by the scope of this book being on international law, the subjects of which are states.[22]

A second issue raised by the harm avoidance principle is whether it would apply in all circumstances or only where the cost of harm is greater than the cost of refraining from harm. I would argue that as a *minimum*, the obligation to refrain from causing harm applies where the cost of refraining from harm is manageable and significantly less than the cost of inflicting harm. This flows from the idea that built into the responsibility for harm principle is a comparison of the state of affairs that is likely to occur without the harmful activity (Pogge 2008: 19–21). Activities that are more harmful than would have been the case are captured by the principle. Economists have generally taken the position that the long-term costs of unmitigated climate change far exceed the cost of mitigation action (Stern 2007).

Thirdly, the harm avoidance principle refers to a 'high probability' of causing harm. How do we define this level of probability? As we have seen after considering climate change science, it is likely or very likely that significant damage is and will occur as a result of anthropogenic climate change (see Chapter 1). While it may be difficult to define in the proposed harm avoidance principle precisely what 'high probability of causing serious harm' means, there is no doubt that projected climate-change-related damage would fall within this principle.

Fourthly, while the harm avoidance principle can be formulated in a manner which does not include reference to human rights, my formulation – by including harm to 'core human rights' – has the advantage that the principle can be seen as a logical extension of the notion of core human rights, which we have seen reflect widely shared values (2.1). By 'core human rights' I am referring to the human rights to life, health and subsistence (see 2.3 below). These rights reflect basic interests essential for human dignity. While some question whether future persons can possess human rights (2.3.3), this is not an obstacle to the harm avoidance principle, which extends to foreseeable harm to the human rights of 'some future persons' upon being born and whatever their identity (Vanderheiden 2008: 137).

A further consideration in applying this principle to climate change harms is that it assumes that only harms caused by human beings are covered. Thus, only anthropogenic impacts would fall within this principle. This is important, given that climate-change-related damage, for example storm surges, may have a combination of natural and anthropogenic causes.

The harm avoidance principle has an important link to other ethical principles which justify climate change mitigation. One such principle – further considered in Chapter 3 – is the polluter pays principle (PPP), which provides that 'those who have caused a problem (such as pollution) should foot the bill' (Caney 2005 [2010]: 125). Thus the PPP entails application of the harm avoidance principle in an environment context.

The harm avoidance principle is also reflected in the 'precautionary' and 'preventative' principles. According to the former, a lack of scientific certainty should not be used as a reason for postponing environmental policies where there are threats of serious or irreversible damage (4.2.6 below).

2.2.1 Objections to a Harm Avoidance Principle

A possible objection to the harm avoidance principle is that it presumes the impossibility of a future technological breakthrough which allows the production of energy which drastically reduces GHG emissions. This objection, however, fails as it ignores the fact that the harm avoidance principle can only be applied based on foresight and knowledge *now*. An ethically justifiable decision cannot retrospectively become unjustifiable because of events which were not foreseeable at the time (Vanderheiden 2008: 186). Moreover, according to scientists, we are simply running out of time to wait for a magical technological silver bullet for the climate change challenge. Experts have pointed out the availability of existing technologies that – given the right regulatory settings – could address the problem (Philibert 2004). Economists applying a cost–benefit analysis have argued that deeper cuts undertaken now will mean a more manageable burden for the next generation (Stern 2007: 284). This is not to deny that mitigation action itself is not without significant costs. As argued above (section 2.2), the harm avoidance principle applies where the costs of not taking action exceed the cost of action and the costs of action are manageable.[23]

Should we reframe our harm avoidance principle to include harm to other species and ecosystems? It is unnecessary to provide an additional justification for climate change mitigation action on this ground as it is sufficient to rely on the impact on human beings to justify climate change mitigation (see below 2.12).

A harm avoidance principle has been criticised for being 'either obscure or overly conservative when taken literally' (Gardiner (2004) [2010]: 13). A related complaint is that the principle is inherently problematic as it involves indeterminacy as to what constitutes harm. But the same criticism can be made of comparable ethical principles such as the proscriptions of lying or stealing. Moreover, there are legal principles with the same level of inherent indeterminacy which does not seem to prevent them functioning. An example is section 23A of the Environment Management and Pollution Control Act 1994 (Tasmania), which provides in section 1 that 'a person must take such steps as are practicable or reasonable to prevent or minimize environmental harm or environmental nuisance caused, or likely to be caused, by an activity conducted by that person'.[24]

Related to this criticism is the notion that the harm avoidance principle is too weak in its content. Put differently, the harm principle might provide grounds for reductions in emissions in GHGs, but this does not take us very far given that the big *distributional* issues of how quickly such reductions should occur and who should be responsible for such reductions remain on the table.[25] There is strength in this criticism in that to address the key climate change policy issues relating to future generations, the harm avoidance principle does not take us far enough: other distributional justice principles must be invoked, including the 'polluter pays' and 'capacity to pay' principles. However, the harm avoidance principle should nevertheless be retained as it provides a basic guidance principle with resonance in the widely shared values which underlie core human rights (see above 2.1).

2.2.2 Individual and Collective Responsibility

The harm avoidance principle outlined above rests on a model of individual responsibility. According to Feinberg (1966: 33, in Hiskes 2009: 43), a person is morally responsible for harm where, firstly, the person caused the harm, secondly, the person was in 'control of events' and they did not occur simply by chance or some external factor, and thirdly, that the first two determinations can be read off the facts or deduced from them. Environmental pollution is difficult to reconcile with this model. While major polluting factories can be identified, 'environmental degradation is also a product of countless everyday decisions of all citizens' (Hiskes 2009: 42). Moreover, some of these decisions 'are not even conscious decisions, and many of them have histories laden with collective policy choices and social traditions' (Hiskes 2009: 42). This applies to climate change, where the range of sources of GHG emissions is extremely wide. Hiskes (2009: 38) argues that in spite of this diffuse and cumulative nature of the harm involved, a harm–rights approach is still feasible. As he puts it:

> it might be true that no one possesses a right-based claim against any *particular* persons to the very real harm of polluting or smoking, but one might still claim a right not to be subjected to such risks or harms. That right would be an emergent right – a protective claim against a harm itself emergent in nature (Hiskes 2009: 38).

The *responsibility* relationship involved in such environmental rights involves inevitably a weaker notion of causality than in Feinberg's scheme (Hiskes 2009: 42). Moreover, the responsibility has a *collective* aspect in that the participation in groups is an essential element in causing the environmental harm involved (Hiskes 2009: 45).

Hiskes' argument that the collective aspect of responsibility for environmental harms cannot be a reason for negating responsibility for harms that violate individual rights is compelling. As he points out, if the contrary were true there could be no responsibility or rights 'in the face of mass behaviour such as panic or war' (Hiskes 2009: 45).

Thus the cumulative and diffuse nature of the sources of GHG emissions do not preclude ethical responsibility for harm. The degree to which individuals, whether acting as consumers, or as directors of companies, or leaders of governments, are responsible must logically be linked to the degree to which they have control or participate in the production of GHG emissions. There is both collective and individual responsibility.

2.3 HUMAN RIGHTS

We have seen that an ethical obligation towards future generations in the climate change context can be based on a harm avoidance principle. Thus, an ethical obligation towards future generations is not dependent on a rights-based approach (Tremmel 2009: 49). As noted above, a harm avoidance principle can be framed without reference to human rights. Why then look to human rights as a source of ethical obligations? The reason is that human rights discourse can potentially add political weight to ethical claims. Moreover, as seen above (2.1), human rights are universal in various ways, with strong support from diverse belief systems and world religions.

It is worth pointing out that in the climate change context, protecting basic rights of persons alive now may go a long way towards protecting future generations (Bell 2011: 104). Given that there will be significant numbers of people alive in 2014 who will be still alive in 2050 or even 2100, the obligation to protect the interests – and rights – of these persons may substantially coincide with what is required to meet the rights of those future generations not yet born (Bell 2011: 104).

A strong argument can be made, however, that beyond 'piggy-backing' on the rights of current generations, future generations have their own rights involving corresponding duties on current generations.

2.3.1 Interest-based Approach to Human Rights

My approach follows Caney (2010a: 166), Bell (2011: 100) and Shue (2011) in arguing that there are certain core human rights – to life (physical security), subsistence and health – which can be derived from especially important human interests.[26] These rights are threatened by climate change, creating corresponding duties of mitigation. Thus climate change threatens: (1) *the right to life*, by storm surges and heat stress leading to human deaths; (2) *the right to health*, by increased incidence of certain diseases; and (3) *the right to subsistence*, through impacts on agricultural production. Human rights provide the strongest moral protection for the key interests listed above (Bell 2011: 103),[27] with rights correlating to duties not to harm these interests.

This approach relies on an interest-based theory of rights: 'other things being equal, an aspect of X's wellbeing (his/her interest) is a sufficient reason for holding some other person(s) to be under a duty' (Raz 1984: 183).[28] These rights are moral rights. At one level their legitimacy is reflected in their embodiment in international human rights instruments. At a deeper level, we saw above (2.1) that their legitimacy is based on a widely shared value that respecting these core human rights is seen as a precondition for ensuring respect for human dignity.

Prima facie climate change would appear to threaten a much broader range of human rights than the three basic rights (of life, subsistence and health) listed above. Caney (2009: 167) has also maintained that climate change impacts the *right to property* through destruction of privately and publicly owned property through increased storm surges and extreme weather events and also the '*right not to be subject to enforced relocation*' given sea-level rise will force inhabitants of low-lying island states and coastal regions to be displaced. My approach follows Caney's approach in (2009 [2010a]) in focusing on a leaner list of three core rights (of life, subsistence and health) for five reasons. Firstly, this avoids reliance on more controversial human rights and thus maximises the likelihood of generating agreement (Caney 2009 [2010a]: 169). This approach is supported by the 'relative universality' of human rights as set out in 2.1 above and reduces the risk of particular rights being identified as being imposed by Western countries. Secondly, it is unnecessary to go beyond these three basic rights to establish an obligation towards future generations in relation to climate change mitigation. This is not to deny that such an approach would be viable. Rather, it is to indicate that this is unnecessary for our purposes: threats to the core human rights are sufficient to generate ethical obligations towards future generations. Thirdly, reliance on these basic rights avoids linkage to a particular political model such as liberalism (Caney 2010a: 164). Fourthly, the approach taken here accepts Shue's criterion for identifying which rights are 'basic'. Thus he identifies basic rights as rights 'the enjoyment of which is essential to the enjoyment of all other rights' (Shue 1980: 19). While not stated explicitly by Shue, a right to property would not qualify as a basic right, as it is not a precondition for the enjoyment of other rights, but rather a vehicle for implementing the more basic right to subsistence (see Shue 1980: 24, 125). If one does not have life or the basics of food and shelter, one does not have the capacity to exercise other rights such as freedom of expression (Campbell 2006: 159). Fifthly, and finally, while a longer list of basic human rights would arguably contain elements of political liberty (for example Shue 1980: 70) or democratic participation (for example Buchanan 2004: 129), these rights are not threatened – or at least not directly – by climate change.

The human right to subsistence reflects universal human needs in that all human beings need food, water and shelter to survive (Campbell 2006: 160). This basic need extends into the future based on a reasonable assumption that the well-being of future generations will also depend on meeting their basic needs for food, water and shelter.

A similar argument can be made in relation to a right to a clean environment. Meeting core human needs for food and water and shelter is also not possible without a minimum level of environmental quality. Without clean air, water and soil, other rights make no sense (Hiskes 2009: 39).[29] Thus environmental harms to air, water and soil give rise to individual – and collective – rights of future generations because the harms from which individuals should be protected are core (Hiskes 2009: 45). Thus it is plausible that while only found in some international instruments (Sands and Peel 2012: 778), from an ethical point of view, a 'right to a healthy environment' can be considered a corollary to protection of the core human rights to life, subsistence and health outlined above. Indeed, Hayward (2005: 48) points out that the 'right to an adequate environment' would protect threats to human interests which are of equal moral importance when compared with other human rights.

Similarly, a 'right to a healthy environment' can be extended to include the 'right to a stable climate' on the basis that the enjoyment of basic rights is contingent upon having a stable climatic system.[30] Thus Vanderheiden argues that 'owing to the critical importance to human welfare of climate stability', a strong case can be made for a 'right to an adequate environment', including 'a claim to climatic stability' (Vanderheiden 2008: 252).

Thus fundamental interests in life, subsistence and health – and a stable climate system – are shared by all human beings regardless of when they happen to be born (Caney 2009: 169). On this basis, future generations have a right to a world free from harm to their basic interests as a result of action by previous generations, with a corresponding duty on current generations not to take action which harms the basic or core interests of future generations (Page 2006: 143–4). Anthropogenic climate change infringes these basic rights to life, subsistence and health. It is the possibility of adversely impacting the interests of future persons by action or inaction now which creates future rights for future persons and corresponding obligations now (Bell 2011: 105).[31] While a 'right to a healthy environment', embracing a 'right to a stable climate system', comprises logical extensions of the core human rights to life, health and subsistence, to date such environmental rights are not well grounded in international instruments (Hayward 2005: 54, Lawrence 2012: 33). Thus, my argument on ethical obligations towards future generations rests upon the less controversial and widely supported core human rights to life, subsistence and health. This is not to deny that an approach resting on a human right to a healthy environment would not be a valid approach.

2.3.2 Claims-based Approach to Human Rights

The argument set out above relies on an interest-based conception of human rights. Can a similar argument be made resting on a so-called 'claims-based' notion of human rights? Under a claims-based notion of rights, 'a person has a right when they are in a position to claim the performance of a duty from another, or to waive it, and therefore to determine by his/her choice how the other ought to act' (Waldron 1990: 95).[32] An objection to the claims-based notion of rights applying to future persons is that future persons are contingent: how can 'possible persons' not yet in existence possess human rights (Bell 2011: 106)? In response to this objection it has been pointed out that the existence of future persons is 'virtually certain' in the sense that, 'barring catastrophe', future persons will exist (Partridge 1990: 53).[33]

Even if one rejects the interest-based notion of human rights and supports the claims-based approach, it can be argued that there is an obligation to avoid action which will very likely result in harm to future persons upon their being born; thus rights can apply to future generations in a *contingent* manner. And thus there is an obligation to take strong climate mitigation action, as failure to take such action will result in the likely infringement of the human rights of significant numbers of people upon their being born (Vanderheiden 2008: 132).[34]

2.3.3 Objections to Human Rights Approaches

(a) Can future generations have rights?

An objection that has been made to ethical obligations to future generations is that there cannot be an existing duty based on a right that does not yet exist and future persons cannot have rights now. Derek Bell has argued that this paradox is not really a paradox at all because 'all human rights-based duties are grounded in the future rights of persons living in the future (even if it is the very near or immediate future)' (Bell 2011: 107). Put differently, whether the victim of the harm in question was alive at the time the duty arose is insignificant in terms of whether a duty exists or not. This is because all duties correlating to human rights relate to persons living in the future (Bell 2011: 108). There is no special problem relating to future generations possessing rights, as the impact on future interests which gives rise to the duties always occurs *after* the agent has taken harmful action, even though it is the *potential* impact which gives rise to the normative duty in the first place (Bell 2011: 108). Thus the apparent paradox involving future human rights giving rise to current duties evaporates. This accords with our intuitive sense. Elliot presents the example of a trap set up in a way which will seriously injure a person upon opening it. If somebody not alive now, in the future triggers the trap, the action taken *now* in setting the trap violates the right of that person. It is morally irrelevant if the victim is alive at the time the trap was set or was born after the trap was set (Elliot 1989: 162).[35]

(b) Non-identity objection

A further obstacle is the so-called 'Parfit's dilemma' (Parfit 1987: 357). Applied to climate change, the dilemma is as follows. If climate change mitigation action is taken, this will change the society in ways which result in different couples coming together and reproducing than would otherwise have occurred. The consequence is that some persons will not be born who would have been born if no climate mitigation action had taken place. According to this view, as it is better to be born – even in a world with climate change – than not to be born at all, the climate mitigation action cannot be justified (Broome 1992: 33–4). As Bell puts it:

> We do not make anyone worse off than they would have been by adopting a business-as-usual policy because anyone who suffers the effects of anthropogenic climate change would never have existed if we had adopted an emissions reduction policy (Bell 2011: 109).

Bell points out that this argument relies on a counterfactual notion of harm, according to which harm occurs when a person's situation is '*worse than it would otherwise have been*' (Bell 2011: 109, italics in original). Bell argues that human rights violations involve harm in this sense, but more fundamentally should be understood as involving a threshold notion of harm. According to this threshold notion of harm, a person's human rights are violated if action is taken which results in a person being left below a threshold required by the right. He gives the example of the emission of GHGs causing storm surges, which violates a future person's right to physical security (Bell 2011: 109).

The non-identity problem does not invalidate this argument for a number of reasons. Firstly, the duty not to violate the human rights of future persons is not dependent on the identity of the rights-bearer (Bell 2011: 109–10). This accords with our intuition. If a person throws a bomb into a marketplace, the identity of the victims is irrelevant to the obligation and rights involved (Lawrence 2012: 39).

Secondly, 'current persons have a duty not to take actions that will violate the rights of the actual future persons who will exist – even if those particular future persons would not have existed but for those very actions' (Bell 2011: 110).[36] This also strongly accords with our intuition. Imagine a small society which releases toxic chemicals into the water supply which slowly poisons the population, with deaths starting to occur in ten years' time. This society considers a 'pollution control plan' which will end the particular industrial practice which is causing the harmful pollution. The 'pollution control plan' will over time cause different persons to be born than would otherwise have been born. But the obligation to take action to avoid harming future persons is owed to all future persons within the scope of the obligation and regardless of who they happen to be and regardless of whether different persons would have actually been born had other policies been in place. The scope of the obligation to avoid harming future persons here does not depend on the identity of the victims. Rather, the obligation is linked to the foreseeability and knowledge of the harm occurring.[37]

(c) Objection of assumptions about future values

What about the objection that a right of future generations implies that contemporaries (arrogantly) make invalid assumptions about the future values and preferences of persons yet to be born. This objection is flawed as we can reasonably assume that the basic interests of future persons – including their need for food, water, shelter and clean air – will be similar to ours (Vanderheiden 2008: 129). Extending this to the climate change context, it seems reasonable to assume that future generations would wish to have as a *minimum* access to a climate which can sustain life at a reasonable standard.

(d) Climate change harms and human rights

We have seen that regardless of which concept of human rights is adopted, prima facie a human rights approach can ground an obligation towards future generations in relation to climate change. However, does the peculiar nature of climate change harms mean that a human rights approach is unworkable? More specifically, does it matter that the harms involved in climate change are from a range of sources – from individuals to corporations – and secondly, that the harm is of a *cumulative* nature. Hiskes points out that human rights are dynamic and must be responsive to emergent types of harms, including environmental harms (Hiskes 2009: 38). Thus human rights must adapt to the new ways in which harm occurs[38] – the novelty of the mode of harm does not make it any less harmful (Hiskes 2009: 46) nor does it weaken the imperative to utilise human rights as a basis of a duty to prevent harm.

Given GHG emissions involve an increased *risk* that various human rights will be violated, is a human rights approach flawed in that the link between the relevant duty and harm is too remote? It has been pointed out that the obligation to promote effective institutions that increase the likelihood of the enjoyment of core rights – and lessen the risk of violation of human rights – can be applied with equal force to traditional civil and political human rights as well as to the threats posed by climate change (Bell 2011: 111–12). Thus we can say that a duty not to torture still has normative existence if we put in place institutions to minimise the risk of torture occurring, even if – in spite of these efforts – it does occasionally occur (Bell 2011: 118). Thus Bell argues for a 'general duty' to promote 'effective institutions that will specify and allocate the more detailed duties needed to ensure the protection of basic human rights from the effects of anthropogenic climate change' (Bell 2011: 114). The consequent allocation of specific duties to individuals – and presumably also to other relevant agents – is a matter of justice (Bell 2011: 114). The daunting nature of this challenge of allocating responsibility in no way undermines the internal coherence of human rights as a basis of obligations towards future generations. There is a parallel here with the harm avoidance principle which can ground a general obligation towards future generations but which needs to be supplemented by distributional justice principles.

2.3.4 Collective Human Rights

The analysis so far has been on the basis of individual rights. Does a similar argument apply to group or collective rights? This discussion is relevant for two key reasons. First, given the link between Western political liberalism and individual rights, a collective rights approach may potentially provide a broader platform which includes cultures, such as those in Asia, which emphasise collective or group responsibilities (Bell 2000: 192–3). Secondly, climate change threatens collective interests such as the destruction of the cultures of low-lying small island states (Bell 2004: 135–52). Does such harm involve damage to the collective rights of a group which is qualitatively worse than the harm occurring to the individuals involved (Bell 2004: 135–52)? If so, this would suggest that there is a particularly strong obligation to avoid harm to the future collective rights of 'peoples' or particular collective groups.

On one view, collective rights constitute no more than the sum of the rights of the individuals involved. The more convincing approach, however, is the 'holistic' view, whereby harm to collective interests involves harm beyond that of the sum of the individual harms (Tremmel 2009: 61–2). The latter is consistent with our intuitive view that genocide – the destruction of an entire ethnic group – is a more grave ethical harm than mass human rights violations (Page 2006: 157). Following on from this point, does the potential harm to particular collective interests in the future justify duties to avoid such harm which are different from duties owed to future individuals?

Brown Weiss argues that future generations can claim a collective 'planetary' right (Brown Weiss 1989: 96). According to this approach, 'generations' hold these rights as groups in relation to other generations – 'past, present, and future' (Brown Weiss 1989: 96). The content of these group rights comprises a right to 'planetary conditions of diversity and quality comparable to those enjoyed by previous generations' (Brown

Weiss 1989: 96). These planetary rights extend to the 'common patrimony', that is, elements of the natural environment such as air, which cannot be assigned to individuals (Brown Weiss 1989: 98–9).

Brown Weiss's notion of each generation possessing a group right seems questionable for the reason that each future generation is not sufficiently delineated as a group, given the reality of overlapping generations (Page 2006: 155). Moreover, in the context of climate change, the harms experienced within future generations may be varied. Indeed, some groups will experience in the short term benefits from climate change, through for example increased food production, while other groups face great hardship, such that it makes little sense to lump all future generations together (IPCC 2007a: 5.8.1).[39]

While future generations may not possess a single 'planetary right', this does not detract from the idea that potential harm to particular collective interests – for example in sustaining a particular island country's culture – can ground especially strong ethical obligations on current generations to avoid action which threatens these collective interests and associated rights (Page 2006: 155–6). Indigenous peoples, who are already amongst the world's most disadvantaged people, are particularly vulnerable to the impacts of climate change owing to their often close relationship to fragile ecosystems threatened by climate change (Cameron 2009–10: 688).

One such possible group right threatened by climate change is the right to development, given that climate change impacts will disproportionately fall on the global poor (UNDP 2007). The right to development is included in a number of international instruments and has been proposed as a collective right of 'peoples' (Shaw 2008: 301–2). The right to development has been linked with intergenerational justice in the Rio Declaration on Environment and Development 1992, which stipulates that 'the right to development must be fulfilled so as to equitably meet developmental and environmental needs of present and future generations'.[40] Similarly, 'intergenerational justice' – or 'intergenerational equity' as international lawyers call it – is part of the concept of 'sustainable development' defined as 'development that meets the needs of the present without compromising the ability of future generations to meet their own needs' (World Commission on Environment and Development 1987: 43). The UN Framework Convention on Climate Change (UNFCCC) incorporates the principle of 'sustainable development', which is to guide parties to the Convention in its implementation, with parties to the Convention having a 'right' and a duty to promote sustainable development.[41] While the norm of 'sustainable development' is part of the UNFCCC, the status of 'sustainable development' and the 'right to development' under general international law remains unclear (Cordonier Segger 2008: 114, see Chapter 4).[42]

Nevertheless, it can be argued that as a minimum, a failure to mitigate climate change threatens the collective *moral* right of future generations to sustainable development as states or peoples, giving rise to a compelling ethical obligation on the current generation to take strong action to mitigate climate change. It is essential to note at this point that establishment of sustainable development as a collective right of this nature is not essential for my thesis, which relies on an obligation towards future generations which flows from the less controversial rights to life, health and subsistence.

We have seen that it is coherent and consistent to adopt a human rights approach to ground an ethical obligation towards future generations in relation to anthropogenic climate change impacts. Such an approach is preferably based on an interest-based notion of rights, with a reasonable assumption that future generations will share current generations' basic interests, the satisfaction of which is dependent on a stable climate. Moreover, even on the basis of a claims-based notion of rights, it makes sense to ground an obligation towards future generations on a *contingent* view of rights. In addition, given that the cumulative nature of climate change impacts and the link between the duty and harms being based on a 'risk', in no way undermines the application of human rights as a basis for ethical duties. While the notion of future generations possessing a single collective right is questionable, it is coherent to conceive of particular groups possessing collective rights in the future – in relation to, for example, culture and sustainable development – which are threatened by a failure to mitigate climate change now, giving rise to particularly strong mitigation obligations on current generations.

2.4 A CAPABILITIES APPROACH

We have seen that human rights provide a basis for action in mitigating climate change for the benefit of future generations. But it needs to be supplemented by distributive justice principles to address the policy issue of how to balance the interests of current and future generations. Nussbaum (2006) and Sen's (2009) capabilities approach to justice has received much attention in recent decades. It has at its core the notion that narrow economic welfare utility – by aggregating welfare, based, for example, on GDP – glosses over inequalities arising from various constraints. Instead, they argue that the focus should be on equalising a set of core capabilities which respect an individual's autonomy to develop in the way in which they choose to do so (Nussbuam 2006: 76–8).[43] These capabilities are: (1) life – or being able to live a life of normal length; (2) bodily health – including food and shelter; (3) bodily integrity; (4) senses, imagination and thought; (5) emotions or attachment; (6) practical reason; (7) affiliation and social interaction; (8) living with other species and living with the concerns for the natural world; (9) play; and (10) control over one's environment (Nussbaum 2006: 76–8).

How does the capabilities approach deal with the intergenerational impact of climate change? Holland has pointed out that enjoyment of the capabilities listed by Nussbaum requires a functioning environment to ensure access to food, fresh water and energy (Holland 2008: 323). This dependency leads Holland to suggest a 'meta-capacity' which she calls 'sustainable ecological capacity', which involves having the environmental resources and services in place to enable the current generation to meet its capabilities 'now and in the future' (Holland 2008: 324). Schlosberg has pointed out that this extended version of the capabilities approach deals well with the intergenerational aspect of climate change. It is unjust to not take strong mitigation action in relation to climate change as climate change will impair the ability of future generations to fulfil their capabilities and flourish (Schlosberg and Ahern 2009). The capabilities approach is preferable to narrower approaches based on, for example, monetary wealth, as it includes social affiliations that will be torn apart for climate refugees (Schlosberg and Ahern 2009: 10).

Page points out that an extended capabilities approach avoids the difficulties thrown up by applying 'welfarism' to issues of intergenerational justice, where welfare is defined as satisfaction of 'desires' or 'preferences'. He points out that a difficulty with 'welfare egalitarianism' is that, if, for example, future generations learned over time to adapt their desires and became '"contented victims" of climate change', welfare egalitarianism would not recognise this is as posing any issue of intergenerational justice (Page 2007: 455).[44] However, a weakness in the capabilities approach is that all capabilities are given equal value, which, as Edward Page notes, is intuitively implausible and also makes for difficulties in making comparisons between persons (Page 2007: 467). Thus, he points out that:

> the flexibility created by refusing to make trade-offs between capabilities below some threshold of minimum functioning also means that, as it stands, Nussbaum's list implausibly attributes equal value to each of the 10 central functioning capabilities from 'life' and 'bodily health' through to 'play' and 'a relationship with another species' (Page 2007: 467).[45]

As a consequence, while a capabilities approach provides a good basis for an overarching obligation on current generations to take climate change mitigation action in order to preserve the capabilities of future generations, it would require considerable further development in order for it to provide guidance on how to address policy decisions involving conflicts between the interests of current and future generations. A difficult issue which would need to be addressed would be how to translate a capability framework into concrete GHG emission reduction targets, given the difficulty of measuring capabilities (Robeyns 2011: section 5).

Indeed, how important is the matrix of justice – the goods with respect to which justice applies – in establishing obligations towards future generations in relation to climate change? Arguably, how we define 'basic goods' under various theories of justice is not an obstacle to justifying an obligation on current generations to take mitigation action on climate change to ensure the protection of future generations. This is because the impacts of climate change for future generations are likely to be so damaging that whatever matrix is adopted – whether human rights, capabilities approach or a needs-based approach – there is an ethical obligation on current generations to avoid such damage.[46]

In this section I have touched on the capabilities approach to justice. Other theories of justice are based on notions of reciprocity, or a social contract, and have also been highly influential, and to these I now turn.

2.5 RECIPROCITY

2.5.1 Direct Reciprocity

'Why should I do anything for posterity? What has posterity ever done for me?' (Narveson 1978: 38)[47] captures the one-way relationship between current generations and our successors. Our successors cannot confer benefits on those alive today. On the other hand, we are in a position to benefit or harm our successors. Thomas Hobbes (1588–1679) took the view that human beings were inherently immoral and that human beings could protect themselves from harm by others by entering into contracts with those others who had the potential to cause harm. In this sense, Hobbes created the concept of 'justice as reciprocity as a balance of deterrence' (Tremmel 2009: 183). Following in this tradition, Barry (1989: 8) regards 'justice as mutual advantage'. Theories of justice resting on a contractual or reciprocal basis (such as Rawls, Gauthier, Barry or Nozik) cannot establish a basis of an obligation towards future generations based on justice.[48] The reason for this is that these theories involve *direct reciprocity* in the sense that justice requires the possibility of a mutual exchange of advantages. This 'tit-for-tat' reciprocity by definition precludes obligations to future generations who cannot offer anything to those alive today.[49]

2.5.2 Indirect Reciprocity

One response to the justice as reciprocity limitation outlined above is to argue that these theories do not establish a basis to ethical conduct at all (Tremmel 2009: 193).

Another response is to stretch the notion of reciprocity to include *indirect* reciprocity. This notion is well captured in the Talmud: 'An old man is asked why he is planting a carob tree, as after all, he will not live to see this tree bloom. He answers: "When I was born the world was full of blooming carob trees"' (Talmud, Ta'anit [2007]: 23).[50] The essence of indirect reciprocity is that each generation has received benefits from its forebears, for example, a good education. The forebears are no longer alive but in return for these benefits the current generation can convey to its successor generation the equivalent benefits, for example, a good education (Becker 1986, in Page 2006: 121). The successor generation can then convey an equivalent benefit to their successors, creating a 'cascade like obligation' (Hoesle 1997: 809, in Tremmel 2009: 195) also characterised as a 'chain of concern model' (Page 2006: 115). Thus human beings' sentimental concern for the well-being of their nearest descendants constitutes a public good which requires every member of society – even those without children – to contribute to its upkeep (Page 2006: 115).

This concept underpins Rawls' 'just saving' principle, according to which:

> Each [generation] passes on to the next a fair equivalent in real capital as defined by a just savings principle ... This equivalent is in return for what is received from previous generations that enables the later ones to enjoy a better life in a more just society (Rawls 1971: 288).

Indirect reciprocity has been combined with a stewardship model by a number of philosophers (for example Brown Weiss 1989: 45; see 2.9 below).

A serious objection to indirect reciprocity is that obligations arising from reciprocity are generally only considered valid if the particular goods were received voluntarily (Page 2006: 123). A newborn child is not in a position to refuse the heritage of humankind or even understand what this would mean.[51]

A further issue is whether indirect reciprocity applies in relation to goods, the consumption of which may be indispensable for survival in the short term.[52] The indirect reciprocity argument seems insensitive to the specific conditions in which each generation finds itself. Indirect reciprocity makes sense in a context where a society is sufficiently wealthy to preserve certain goods such as education, but what if the society is poor? Can one a priori make assumptions about what is appropriate to convey to the next generation? What if the society is under such pressure to survive that consuming resources now is the only way of stopping poor people dying right now? In such a context it would seem unethical for the current generation to sacrifice their interests to those of the future as this would mean an arbitrary decision to favour the interests of remote future generations over the interests of those alive today (see Chapter 3).

Moreover, the indirect reciprocity argument seems intertwined with an assumption of progress. Surely for some generations born at particular times – or in the future – they could be in a situation where – on balance – they understandably take the view that their inheritance has more negatives than positives. Of course, this begs the question of comparison with what? Yet human beings have a strong survival instinct to make the best of difficult situations. Artificially constructing a notion of consent to the benefits and burdens of one's inheritance seems a distorted characterisation of this reality.

In summary, then, justice theories relying on direct reciprocity cannot establish a basis for an obligation towards future generations. Indirect reciprocity also runs into difficulties as one's inheritance is not voluntarily received. Moreover, indirect reciprocity seems to be insensitive to the particular situation each generation finds itself in.

2.6 JUSTICE AS IMPARTIALITY

Many of the most influential theories of justice are derived from a notion of impartiality. The basic idea here is that justice is derived from an agreement reached by rational people in a context where they are unable to translate bargaining power into advantage but rather are able to find a basis for agreement 'acceptable from all points of views' (Barry 1989: 7). One of the most influential of such theories was developed by John Rawls (1971).

2.6.1 Rawls

The essence of Rawls' theory was to conduct a thought experiment according to which rational persons guided by self-interest would come together to decide on just principles. The thought experiment would occur under a so-called 'veil of ignorance' according to which participants would not know whether they were to be rich or poor, intelligent or dull. Rawls argues that the participants would agree to principles according to which inequalities would be allowable only to the extent that the least advantaged enjoyed the greatest benefit ('difference principle') as any participant could be born into this group. Inequalities would still be allowable to the extent that inequality would further economic growth with trickle-down effects to all. Rawls recognised that this created a problem in terms of justice between generations, as people in the original position could decide to favour their generation at the expense of their successor generations (Rawls 1971: 140). He therefore extended the thought experiment so that the representatives of past, present and future generations were represented in the original position (Rawls 1971: 288).[53]

Rawls' next stage of the argument is to maintain that in the original position participants would agree on a so-called 'just savings' rate according to which 'persons in the original position are to ask themselves how much they would be willing to save at each stage of advance on the assumption that all other generations are to save at the same rates' (Rawls 1971: 287). Rawls recognised that the just savings principle would entail a savings rate which changed in accordance with the level of development of society. Thus he states that '[w]hen people are poor and saving is difficult, a lower rate of saving should be required; whereas in a wealthier society greater savings may reasonably be expected since the burden is less' (Rawls 1971: 287). Rawls includes an indirect reciprocity element by suggesting that each generation pass on a fair equivalent in real capital (including culture and education, not just material capital) which is in return for what they have received (Rawls 1971: 288). He acknowledged, however, that reciprocal exchanges between generations can at most be 'virtual ones', involving adjustments in the original position in drawing up the just savings principle (Rawls 1971: 291). Rawls assumes that each generation cares for its immediate descendants (Rawls 1971: 288), an assumption at odds with his assumption of participants in the original position acting with 'self-interest'. This assumption was dropped in his later reformulation of this theory (Rawls 2001: 160).

Rawls' theory has been criticised for its failure to take into account depletion of non-renewable resources and environmental damage by taking for granted that later generations will be better off owing to technological advantages (Birnbacher 1977: 386–7, in Tremmel 2009: 155). Arguably Rawls' just savings principle is given unwarranted importance in his theory given that historical evidence suggests that significant drops in wealth between generations have occurred due to 'wars, enslavement and depression' rather than decisions about savings (Tremmel 2009: 165).

2.6.2 Tremmel

Taking account of these criticisms, Tremmel (2009: 170) argues that Rawls' construct can be extended to justify the following two principles which would be agreed by the representatives in the thought experiment:

1. Maximise the average individual well-being of all members of all generations. This principle primarily obliges every generation to avoid wars as well as ecological, societal and technological collapses that might significantly impair human welfare.
2. No generation is obliged to save more than the previous generation.

Tremmel extends Rawls' theory by including axiological elements absent from Rawls (namely human welfare). His first principle takes into account ecological threats in a manner absent from Rawls' theory.

Tremmel's variant of Rawls is valuable in providing a principle of higher-level generality. It also meshes well with scientists' concerns of catastrophic climate change impacts flowing from crossing certain thresholds relating to the global climate system (1.1.2 above). However, because of its high-level generality, it arguably does not take us much further in addressing the difficult issue of how to balance conflicting generational interests in the climate change context. For example, in applying Tremmel's general principle which requires every generation to 'avoid ecological collapses' that 'might significantly impair human welfare', it remains unclear how this principle should be applied in a context where sharp GHG emission reductions impede human welfare, at least in the short term. By referring to *the average* individual well-being of all generations being maximised, does this principle allow decisions whereby particular groups' or nations' interests are sacrificed so long as the overall average is increased? Like the non-harm principle espoused above, this broad-brush principle has value as a motivational tool but the hard distributional justice issues remain unresolved.

Tremmel's second principle is close to Rawls' 2001 formulation, which Tremmel (2009: 158) points out is a variant of the Golden Rule: 'do unto others as you would have them do unto you.' But if a broader view of the economy is taken which includes ecological destruction, it is questionable whether a narrow focus on savings makes sense. What is the value of focusing on savings if broader ecological collapse is ignored?

2.6.3 Brown Weiss

Brown Weiss (1989) extends John Rawls' theory of justice in a different direction. The essence of Brown Weiss's (1989: 24) view is that: if each generation did not know beforehand when it will be the living generation, it would rationally choose a principle whereby each generation:

> would want to inherit a common patrimony of the planet in as good condition as it has been for any previous generation and to have as good access to it as previous generations. This requires that each generation pass the planet on in no worse condition than it received it and provide equitable access to its resources and benefits.

This concept has appeal in terms of being a compromise between an extreme 'right of exploitation' and an extreme deep ecology/conservation perspective. Brown Weiss's theory provides flexibility in the management of natural resources: such exploitation is allowed provided that overall the planet is left in no worse condition. Thus Brown Weiss's principle involves conservation of the diversity of the natural and cultural resource base so long as future generations' options are not negated (Brown Weiss 1989: 41–2). This is linked to an 'equality' principle, with Brown Weiss (1989: 24) stating that 'the theory of intergenerational equity calls for a minimum level of equality among generations'.

Critics have pointed out that Rawls/Brown Weiss's approach is limited in its strong anthropocentric bias.[54] It also has built into it a strong concept of progress. Thus, one can legitimately ask why it would not be sufficient for each generation to pass on the earth in sufficient condition to allow future generations to lead a reasonable quality of life (Beckerman 1999: 73).[55]

Can the Rawls/Brown Weiss model provide a principle of justice which would require the present generation to limit its GHG emissions even where the benefits of such actions would largely manifest themselves only in the future? Vanderheiden (2008: 119) argues that applying Rawls' theory is untenable in relation to climate change in that 'ignorant of the stage of national development in which they will live, no persons could endorse a standard zero emissions growth, for this would impose a cap on development for those societies in the early stages of industrialization'. Adopting such a standard now would be unconvincing as it would imply a desire to forgo industrialisation entirely. Similarly, applying Rawls' thought experiment to climate change, it would not make sense to endorse high GHG emissions caps as this would in effect endorse climate change. Vanderheiden (2008: 120) points out that it would only make sense for countries to endorse strict emission cuts if they could be assured that other nations also endorsed similar cuts, otherwise they would suffer competitive disadvantages in development.

But an answer to Vanderheiden is that he is seeking to extract an overly precise policy prescription through application of the thought experiment. If the thought experiment is used to justify a broad principle, according to which each generation puts in place policies to avoid dangerous climate change, the difficulty identified by Vanderheiden evaporates. If one happens to be born in an early pre-industrial generation, there is no need to place caps on GHG emissions. On the other hand, if one is a member of the current generation, there would be an obligation to impose stringent GHG emissions caps. Nevertheless, Vanderheiden's critique remains valid in the sense that the Rawlsian thought experiment cannot take us very far in terms of specific obligations in relation to climate change, as these must be contextual, related to the circumstances applicable in time and place.

This discussion suggests that a deeper problem may lie behind all Rawlsian-based theories in their application to intergenerational justice. There is a great attraction in finding a general principle which gives content to justice in an intergenerational context. But as we have seen, it is impossible to find a principle which involves any precision which is independent of the time period for which it is to apply. To date, Rawlsian theories have reflected notions of progress, with the content of intergenerational justice being, for example, the principle that the earth should be handed over in no worse condition than it was inherited. But if we think ahead to the earth in 10 or 20 years from now, what content could such theories have? If, for example, scientists tell us that we are on the cusp of a dangerous tipping point which could seriously impact the prospects of *all* future generations, surely a narrow reciprocity principle, according to which the generation in power does no more than the feeble efforts of its forebears to avert catastrophe, would be totally inadequate as a basis for action. While Rawls adopted an anthropocentric view, he acknowledged that there was no definite answer in terms of how 'the burden of capital accumulation and of raising the standard civilization and culture is to be shared between generations' (Rawls 1971: 286). On the other hand, he argued for 'certain bounds which impose significant ethical constraints' (Rawls 1971: 286).

2.7 COMMUNITARIAN APPROACHES

Rawlsian theories of intergenerational justice seem to be dependent on a questionable notion of indirect reciprocity. In contrast, communitarian approaches seem to offer a strong argument for obligations to future generations not dependent on reciprocity. According to these views, justice stems from a shared sense of community (for example Sandel 2009). This sense of community comes from a shared past. De-Shalit (1995: 12) has extended this sense of community into the future to create the concept of a 'transgenerational community', a community which extends into the future spanning generations. A community can be defined as a group of people with (1) shared values ('moral similarity'), (2) shared cultural interaction – where 'culture' is defined broadly to include all forms of communication, not just the arts, and (3) interaction in daily life (De-Shalit 1995: 22, 23). One's identity is intertwined with one's community. Obligations to future generations are grounded in rational self-interest as an essential component of preserving one's identity, as one's identity is rooted in the past and continues into the future. Thus De-Shalit (1995: 15–16) argues that:

> … the constitutive community extends over several generations and into the future, and just as many people think of the past as part of what constitutes their 'selves', they do and should regard the *future* as part of their 'selves'. These are the relations that form the transgeneration community, which is a source of our obligations to future generations.

De-Shalit, Taylor and Sandel all acknowledge that their communitarian concept of justice is dependent upon a shared conception of the good that defines that particular community and that community only (Hiskes 2009: 15). These communitarian theories, however, can only provide a basis for an obligation towards future generations of one's own particular community and, on the face of it, theories such as that of De-Shalit 'cannot offer any reasons for people in rich countries to cut back so as to improve the prospects of future people in other communities' (Barry 1999: 99).[56] But on closer analysis, a global community in relation to climate change may be identified (see 2.8 below).

Applying communitarian-based theories of justice to climate change presents difficult problems because, by its very transboundary nature, climate change involves 'the ethical claims of non-compatriots' (Page

2006: 120).[57] Cosmopolitan theories seek to overcome these limitations and it is to these theories we now turn.

2.8 COSMOPOLITAN THEORIES

We have seen that communitarian theories of justice are limited in only providing a basis for obligations towards members of one's own society. In contrast, *cosmopolitan* theories of justice rest on the notion that obligations are owed to all persons regardless of which state they happen to be born into. Cosmopolitan theories of justice contain three core elements: firstly, 'the ultimate units of concern are human beings or persons, rather than tribes, nations, states' (Pogge 2008: 175); secondly, 'universality' with the 'status of the ultimate unit of concern attaching to every living human being equally'; and thirdly, 'generality, the special status has global force, and is owed not just to compatriots' (Pogge 2008: 175). Within cosmopolitan justice theory there are a variety of perspectives, ranging from Peter Singer's (2002) utilitarianism to Simon Caney's (2005) human rights approach (Harris 2010: 102).

Cosmopolitan theories of justice would seem to offer great promise in establishing obligations owed towards future generations as responsibilities are owed to other persons regardless of 'spatial or temporal distance' (Attfield 2003: 162).[58] However, cosmopolitan theories have been criticised for assuming the existence at the international level of a genuine justice community in the sense of globally shared values, identity and institutional structures (Vanderheiden 2008: 97–8). This point certainly holds true in relation to anti-poverty efforts and redistributive economic development, with international law failing to impose obligations in this area (Vanderheiden 2008: 97–8). However, at least *elements* of an international justice community are evident in the discourse on human rights and in treaties aimed at furthering trade liberalisation (Hurrell 2004: 42, in Vanderheiden 2008: 98). Moreover, it could be argued that we only need to demonstrate the existence of a global community in relation to core human rights impacted by climate change. We have seen that these rights reflect widely shared values (2.1).

In addition, the objections to cosmopolitan justice cease to apply in relation to climate change as '[a]ll depend on a stable climate for their wellbeing, all are potentially affected by the actions or policies of others' and none can fully opt out of any effective cooperative scheme to mitigate emissions (Vanderheiden 2008: 104).

Central to cosmopolitan approaches is the idea that causation of harm is sufficient to give rise to obligations in spite of geographic distance between the person causing the harm and the victim. Paul Harris (2010) applies cosmopolitan justice theory to climate change, building on Pogge's argument that we have a moral duty to help those people in whose harm we are materially involved (Pogge 2005: 46). Harris (2010: 7) argues that this obligation should be extended to impose an obligation on all affluent people everywhere to limit their GHG emissions. He argues that this obligation applies not only to wealthy people in industrialised countries but also to the burgeoning affluent middle classes in larger developing countries such as China and India. Harris (2010: 168) argues that this cosmopolitan justice approach should supplement rather than replace the (inadequate) state-based international regime currently being negotiated.

Thus if we consider that there is a global community bound together by a concern for core human rights and the interdependence of all people concerning climate change causes and impacts, the obstacles to communitarian theories – such as that of De-Shalit – can be overcome. Thus an extension of the notion of community across national boundaries and into the future can provide a further basis of an obligation towards future generations in relation to climate change.

Cosmopolitan justice theories have been criticised for failing to indicate how the principles can be implemented in the current global climate architecture, which is based on state sovereignty. For example, Page (2008: 570) states that 'it remains unclear how Caney's methodological individualism can be operationalised given the national focus of current global architecture'. However, cosmopolitans in reply argue that this gets matters the wrong way around. Global architecture should be brought into alignment with cosmopolitan justice principles, not the reverse.[59] Indeed in Chapter 4, I use justice principles to critique the current international climate change regime. While in my view the sovereign state remains likely to be the central component in building an effective international climate change regime, it is essential to see how this system needs to accommodate cosmopolitan justice principles.

Elements of cosmopolitan justice are already part of the global climate change negotiation discourse, for example the concept of equal per capita emissions espoused by India (see Chapter 6). In addition, international climate change negotiations inevitably reflect national climate change policies. In the national climate change policy debate, the concept of equality, which is at the heart of cosmopolitan justice, is likely to play a very significant role, including application to the difficult issues of intergenerational justice.

2.9 TRUSTEESHIP/STEWARDSHIP

Brown Weiss's theory of intergenerational equity includes the idea of a 'trust', with each generation holding the earth 'in trust' for future generations (Brown Weiss 1989: 17). The essence of Brown Weiss's theory is that:

> ... each generation should be required to maintain the quality of the planet so that it is passed on in no worse condition than the present generation received it, and should be entitled to a quality of the planet comparable to the one enjoyed by previous generations (Brown Weiss 1989: 38).

Brown Weiss (1989: 17) argues that human beings 'as a species, hold the natural and cultural environment of our planet in common, both with other members of the present generation and with other generations, past and future'. She goes on to argue that each generation, at any given time, is a 'trustee of the planet for future generations' and derives from this trusteeship an obligation to 'care for the planet' and an obligation towards future generations to pass on the planet in no worse condition.[60]

The concept of a trust has parallels with theories of 'stewardship', which seek to ground intergenerational equity on a religious, particularly Christian basis, where for example, human beings are under a duty to take care of God's creation.[61] A difficulty with Christian arguments in favour of stewardship is that the Bible can also be interpreted as justifying the exploitation of nature (Gillespie 1997: 71–6). A good example is the American Interior Secretary in the early 1980s, James Watt, who 'justified his desire to open up nearly 800 million acres of Federally owned land to immediate corporate exploitation because his "responsibility is to follow the Scriptures which call upon us to occupy the land until Jesus returns"' (Brown 1981, in Gillespie 1997: 73). Without belief in a deity, it is difficult to see on what basis this notion rests. Moreover, conflicting interpretations of the Scriptures present difficulties in grounding intergenerational equity on a religious basis.[62]

It seems to the author that the notion of a trusteeship is a somewhat artificial construct designed to restrain human beings' domain over nature (Lowe 1999: 27). The concept of a 'trust' assumes 'ownership' of the earth's resources in a deity or other entity (Goodin 1985: 175), and therefore rests on a basis upon which agreement may be very difficult to reach. However, this is not to deny that the legal notion of a 'trust' may provide a vehicle for implementing obligations towards future generations in relation to climate change (see 7.6.1).

2.10 SUSTAINABILITY

Does sustainability discourse provide a basis for intergenerational equity?

Vanderheiden (2008: 132–8) seeks to justify an obligation to future generations in the climate change context on the basis of the concept of sustainability. He posits the question of whether 'there exists a moral obligation to manage the atmosphere in a manner consistent with its sustainable use' (Vanderheiden 2008: 133). He argues that a duty of resource conservation may be found through 'consideration of the nature of environmental harms' (Vanderheiden 2008: 133).

He points out that:

> ... for conservation to be effective, a steady commitment to its imperatives over time is required. A law that mandated sustainable forestry practices on every day but Sunday would obviously be ineffective in maintaining a sustainable forest (Vanderheiden 2008: 133).

He argues that conservation imperatives must allow for causal chains which stretch over time (Vanderheiden 2008: 133). This argument seems to be circular in that 'conservation', or 'sustainability', includes within its own definition an obligation which extends over time (Vanderheiden 2008: 132). But it is the basis of this obligation that is the question, so Vanderheiden's argument seems to assume the obligation which he is seeking to establish. While it is correct to say that sustainability and conservation policies, to be 'effective', must stretch out into the future, effectiveness is not what is at question, rather the justification for taking into account the interests of future generations. A normative rule cannot be deduced from an effectiveness argument. Indeed, an obligation to future generations seems to be the basis of sustainability rather than the other way round.[63] Sustainability is arguably the only possible strategy for implementing effectively an obligation towards future generations (see Chapter 7).

2.11 DEEP ECOLOGY AND FUTURE GENERATIONS

A possible objection to the harm avoidance principle and the justice theories discussed so far is their basis in human values: it is harm to *human* interests, not damage to the global ecological climate system, that triggers the harm avoidance obligation and human rights duties. Rawlsian theories aim to preserve the ecological system for the benefit of human beings. And communitarian theories are concerned with the community of human beings. This raises the issue of whether it would be more convincing to base an ethical obligation to address climate change on a 'deep ecology' approach, which entails assigning value to the global ecological system as such, with human beings just one species within this integrated whole.[64]

The notion of 'weak anthropocentrism' – an environmental ethic resting on a human sense of harmony with nature – arguably can provide a sufficient basis to proscribe environmentally destructive behaviour (Norton 2003: 163). Human beings share with other species a common requirement for clean air, water and sustainable ecosystems (Norton 2003: 172). This mutuality of interests means that 'weak anthropocentrism' can ground a 'strong sustainability' approach: long-term human values require protection of the ecological system (Norton 1991: 240).[65] Thus from a practical policy point of view, weak anthropocentrism points generally in the same direction as deep ecology approaches.

This point applies with equal force to climate change impacts: human beings share with other species a dependency on a stable climatic system. So the reason to mitigate climate change to prevent, for example, destruction of the polar bear may overlap with the reasons to protect human beings, who also depend on the climate system.

Is it true, however, that deep ecology and weak anthropocentricism always point in the same direction in relation to climate change policy? One area of possible policy divergence is in relation to plantation forests. An anthropocentric approach may seek to expand plantation forests given their capacity to act as carbon sinks (that is, absorbing CO_2), but this could reduce biological diversity and in this respect harm the global ecological system. But even on this issue, a strong argument can be made for maintaining biological diversity and seed varieties as a large gene pool can assist human beings in adapting to climate change through, for example, developing drought-resistant crops (Food and Agricultural Organization (FAO) 1997). A further point of possible divergence is in relation to climate change adaptation. The fact that climate change has already sped the extinction of various plant and animal species (Maclean and Wilson 2011) may suggest that human beings are more resilient than other animals and therefore not necessarily sharing their interests. But this may be simply an issue of the timescale involved with human beings' long-term survival as a species equally threatened if strong mitigation action does not occur. Climate change science does not give any definite answer to this question (1.1.2 above).

The advantage of arguing from an approach resting on human values is that it avoids thorny problems entailed in deep ecology approaches, including how to decide on priorities in relation to harm to human beings and other species. Deep ecology approaches arguably do not have any coherent account of value in cases of conflict between different species

(Stone 2007: 199). Furthermore, there is no single non-anthropocentric standpoint, with ecocentricism, biocentrism, sentientism in conflict with one another on certain issues (Hayward 2005: 35).[66] But an even more fundamental concern is that deep ecology approaches may be less persuasive to many people, whereas approaches building on traditional anthropocentric theories of justice and ethics may have more chance of gaining traction and ultimately building a global consensus for action on climate change. Andrew Light makes this point in relation to environmental protection generally, arguing for 'practical pluralism' in environmental ethics which seeks to 'create links between already existing moral priorities in specific human communities and the ends of environmental concerns' (Light 2003: 234). He thus draws links between the concept of human beings being part of nature found in a number of cultural and religious systems, including Naess's deep ecology, early Christianity, Buddhism and Spinoza's philosophy (Light 2003: 234).

So the approach taken in this book is human-centred and rooted in a concept of human beings being part of and dependent upon the global ecological system, including the climate system. The approach taken is not hostile to deep ecology approaches but rather sees them – at least in relation to climate change – as pointing in the same policy direction with the hope that by building on existing human-centred ethical and justice theories, strong action on climate change is more likely to become a reality.

2.12 CONCLUSION

This chapter has established that current generations have an ethical obligation towards future generations to take strong mitigation action to avoid anthropogenic harm to the climate system. This approach rests on the premise that 'respect for human dignity' requires respect for core human rights of life, health and subsistence. These core human rights and respect for human dignity comprise widely shared values (2.1). A harm avoidance principle represents the corresponding ethical obligation or duty corresponding to these core human rights. In the context of climate change, this ethical obligation arises given the high probability of current generations' actions impacting on future persons' core human rights. While there are strong arguments for a 'human right to an adequate environment' and a 'human right to a stable climate', to date these moral rights remain controversial. This is not to deny the validity of an approach based on these environment/climate-based rights. The approach taken in this book, however, is to rely upon the less controversial human rights to life, health and subsistence, which are adequate in generating ethical obligations towards future generations.

We saw that objections to the notion of future persons possessing rights are unconvincing as they ignore the fact that ethical obligations are not dependent on knowledge, or the identity, of the actual persons harmed. These approaches also ignored the prospective nature of all human rights duties.

Turning to theories of justice, we saw that Nussbaum and Sen's capabilities approach provides a solid basis for an overarching obligation towards future generations to preserve their capabilities. While the impartiality and equality principles embedded in Rawlsian theories remain appealing, Rawlsian theories remain inadequate to address intergenerational aspects of climate change, as they cannot deliver principles sufficiently detailed and applicable independent of the time context in which policy choices are being considered.

Reciprocity-based justice theories were found to be problematic in grounding an obligation towards future generations owing to the generally one-way relationship between current and future generations. However, communitarian theories – if combined with cosmopolitanism – can provide a basis for an obligation towards future generations on the basis of a global community extending into the future, based on core human rights and the causal interdependency of people in relation to climate change. Crucially, the obligations involved extend across national boundaries.

While the analysis in this chapter has focused on obligations towards future unborn generations, strong climate change mitigation action can also be grounded on the need to take into account the interests of people *alive today*, particularly given the interest of younger people to eventually bring children into the world in a context where their children's future is not imperilled. In this sense the duty to take climate change mitigation action may – at least partly – be based on self-interest given the inextricable link between the interests of parents and grandparents and their children and grandchildren.[67]

This chapter has sought to establish an overarching ethical obligation on those in positions of power to mitigate climate change in order to protect the core interests and human rights of future generations. This obligation entails an ethical obligation on states but nevertheless remains incomplete in a number of respects. The content of the obligation is to take strong mitigation action to ensure that the climate system is not pushed over a threshold which results in significant harm to the core interests and human rights of future generations. But how exactly is this threshold to be defined? This and other important issues remain unresolved. These issues include the rate at which mitigation should occur – entailing balancing the interests of contemporaries and future generations, determining which states have this ethical obligation (for example only developed) and consequent mitigation burden, the relevance of historic emissions and knowledge of emissions. These further issues can only be addressed by application of distributive justice principles, which are taken up in the next chapter.

NOTES

1. This chapter relies on Peter Lawrence, 'Justice and Future Generations: Environment Discourses, International Law and Climate Change' in: Brad Jessup and Kim Rubenstein's (eds) *Environmental Discourses in International and Public Law* (Cambridge University Press, 2012) 23–46.
2. Ross Garnaut (2008c).
3. These rights thus comprise 'moral rights' in the sense that they entail a claim that 'certain social or legal norms ought to exist' (Campbell 2006: 28). Caney uses the notion of human rights as constituting 'moral thresholds' below which persons ought not to be allowed to fall (2010a [2010] 164). Caney's approach of building his argument on a shared and minimal concept of human rights is inspired by Pogge (2008: 51–2), although their methodologies are not the same.
4. My approach is inspired by Moellendorf (2009: 7) who bases a theory of justice to address global inequality on the inherent dignity of persons.
5. The Preambles to both the International Covenant on Civil and Political Rights and the International Covenant on Economic and Social Rights also assert the value of human dignity (Moellendorf 2009a: 6).
6. McCrudden (2008) traces the history of the notion of 'human dignity' and its expression in human rights instruments. The other core element of 'human dignity' involves the notion of 'the state exists for the sake of the individual and not vica-versa [*sic*]', which is not relevant for our purposes (724).
7. Tomuschat (2008: 83) points out that there is disagreement between Western countries on the death penalty, which seems to undermine the idea of the right to life being a universal right. Such disagreement is, however, consistent with a shared notion of the right to life in the sense of a duty to not arbitrarily deprive persons of their life. This core meaning is sufficient for our purposes in relation to grounding ethical obligations in relation to climate change.
8. Renteln (1990: 47) contains a good overview of the various critiques of these conceptual notions of universality.
9. International legal universality is not inconsistent with failures to enforce international human rights norms at the national level (Donnelly 2007). It is also consistent with there being considerable scope for diversity in the implementation of human rights norms in the context of different cultures (Langlois 2001: 102).
10. Susan Waltz (2001) has pointed out, however, that a wide range of countries outside the Western block had a significant impact in the drafting of the Universal Declaration of Human Rights, particularly in the inclusion of economic and social rights and strong non-discrimination principles.
11. Sim focuses on Melden's (1977) theory of rights, which embeds 'rights' in relationships and the moral community (Melden AI, *Rights and Persons* (University of California Press, 1977)).
12. See contrary view of Freeman (2011: 119).

13. Thus Donnelly (2007: 286) argues that human rights have a functional universality. However, the theory of 'modernity' as a universal process entailed in this approach would need to be demonstrated.
14. China has undermined the notion of universality in various ways, including its view that implementation of human rights is dependent on the level of development (Sceats and Breslin 2012: 7, 8).
15. In the 1991 White Paper, the Chinese government asserted that: 'the right to subsistence is the most important of all human rights, without which the other rights are out of the question' (quoted in Sceats and Breslin 2012: 7). China has signed but not yet ratified the ICCPR.
16. Langlois (2001: chapters 1 and 2) demonstrates that the invoking of 'Asian values' by political leaders in Malaysia, Indonesia and Singapore in the 1990s was done in an instrumental way to preserve political power and shut down dissent. However, Langlois's (2001: 70–71) empirical work demonstrates that while there is no such thing as 'Asian values', there are located within Asia ways of conceptualising human rights which are quite different from those prevailing in the West, particularly in relation to the role played by public religion.
17. Following Caney (2010a [2010]: 165) my approach is consistent with the view of those who base human rights on some notion of 'intrinsic value' as well as those who seek to ground human rights on their *functioning* in addressing needs common to all societies such as peacefully resolving conflicts, for example Buchanan (2004: 149).
18. Cf. Hulme (2009), who is sceptical of the possibility of reaching the necessary level of agreement to build an effective climate change regime.
19. Responsibility to avoid harm can only logically be attributed to those with genuine freedom – and power to make decisions which can prevent harm occurring. See a critique of harm avoidance as a basis for obligations in relation to poverty in Allen Buchanan and David Golove (2002: 905).
20. John Stuart Mill, *On Liberty* (1859 [1965]: 126, 135). There are parallels here with Thomas Pogge's negative responsibility for world poverty. See Pogge (1998: 501).
21. Lawrence (2012: 44–5) argues that a harm avoidance principle meshes well with Hart's minimum content of natural law (Hart HLA, *The Concept of Law* (Clarendon Press, Oxford 1994) 193).
22. A similar point is made in relation to the bearers of obligations of distributive justice principles addressed in Chapter 3 (3.2). The notion of ethical responsibilities of governments is quite distinct from the international law doctrine of state responsibility (4.3.2), although the latter clearly reflects the former in its content. The notion of state responsibility, however, goes beyond the 'principle of harm avoidance' argued for here in that it entails obligations of restitution, including compensation where states fail to avoid causing harm. This notion of restorative justice is distinct from my more forward-looking principle of harm avoidance.
23. See 3.5.3 below, which discusses how to resolve conflicts between the rights and interests of contemporary and future generations, particularly the poor.
24. Accessed 12 October 2012 http://epa.tas.gov.au/policy/empca. See also section 28 of the South African National Environmental Management Act 107 of 1998. http://www.elaw.org/node/2702.
25. Allen Buchanan and David Golove (2002: 905) make a similar argument in relation to poverty, arguing that rather than rely on an obligation not to harm, it is preferable to go straight to distributive justice principles.
26. These 'basic rights' are roughly the same as those argued for in Shue (1980).
27. Bell relies on Griffin (Griffin J, *On Human Rights* (Oxford University Press 2008) 35).
28. Waldron (1984: 183) quoted in Page (2006:144).
29. Hiskes uses the term 'basic' rather than 'core' but the meaning is the same.

30. Hayward (2000: 558) points out that more than 30 national constitutions enshrine rights to the environment.
31. Bell relies on Elliot (1989: 162).
32. Quoted in Page (2006: 143).
33. Discussed in Derek Bell (2011: 106).
34. Vanderheiden relies on Feinberg (1966: 139).
35. Quoted in Bell (2011: 107).
36. For rebuttals of the non-identity problem, see Tremmel (2009: 39–46) and Caney (2006: 255, 268).
37. The same argument applies to obligations based on future persons' human rights: see Derek Bell (2011: 109–10).
38. Judicial authority for this is *Tyrer v United Kingdom* (1978) 2 EHRR 1.
39. The Intergovernmental Panel on Climate Change (IPCC) states that 'The preponderance of evidence from models suggests that moderate local increases in temperature (to 3°C) can have small beneficial impacts on major rain-fed crops (maize, wheat, rice) and pastures in mid- to high-latitude regions, but even slight warming in seasonally dry and tropical regions reduces yield.' The report goes on to state that the increased frequency of severe weather events will 'disproportionately impact' smaller-scale farmers (IPCC 2007a: 5.8.1).
40. Principle 3.
41. Article 3.4.
42. Cf. Sands and Peel (2012: 206–17).
43. Sen, however, opposes the idea of a list. See Robeyns (2011: 14–17).
44. Page also canvases deficiencies in 'resourcism', including its implication that one can compensate irreplaceable elements of the global ecological system (457) and that it downplays the heterogeneity of human well-being, with mental illness for example not being reducible to physical resources (459).
45. For a further critique of the capabilities approach, see Page (2006: 67–71).
46. A human needs approach is used in Tremmel (2009: chapter 5).
47. Quoted in Tremmel (2009: 196). (Narveson J, 'Future people and us' in RI Sikora and B Barry (eds), *Obligations to Future Generations* (Temple University Press 1978) 38–60).
48. Hiskes (2009: 48). These authors do have theories about obligations towards future generations but they lie outside the particular theory of justice involved. Rawls does have a theory of just savings upon which such an obligation rests, but this is strictly outside his theory of justice (2.6.1 below).
49. A deeper problem with these theories is that they assume all behaviour is selfish. Moreover, they fail to capture the sense of fairness in common-sense morality: social institutions such as friendship and marriage can't survive on the basis of direct reciprocity. See Hiskes (2009: 52–3).
50. Quoted in Tremmel (2009: 194).
51. Tremmel (2009: 196) seeks to meet this objection by stating, 'But if it were possible to ask an average member of the next generation, what would his answer be? Surely "yes". Almost everything seems to be better than not to exist.' But he goes on to acknowledge Partridge's argument that this involves making comparisons between a certain condition of living and 'never existing', which is invalid (Partridge 2007: 14, in Tremmel 2009: 196) (Partridge E, 'Should we seek a better future?' .*The Online Gadfly* (2007) http://gadfly.igc.org/papers/swsabf.htm (accessed 4 January 2008)). Tremmel (2009: 196) goes on to point out that the average representative of the next generation might in fact say that they would like to accept the positives – democracy and satellite TV, but reject the negatives such as AIDS and poverty. Tremmel's (2009: 196) response is that the so-called 'cherry picking' is unfair just as it would be under private law in the case of a private inheritance which cannot be partly rejected. But surely this is a weak argument as a private inheritance can be completely rejected in a way that is simply not possible with

intergenerational inheritance. It is difficult not to agree with Page (2006: 123) that the lack of voluntariness remains a difficulty in this argument. See further discussion in Page (2006: 121–124).

52. Page touches on this but does not develop it, stating that it is hard to believe that it may be appropriate for present persons to save goods rather than pass them on to future generations 'when the persons concerned are impoverished members of developing countries' (Page 2006: 121).
53. Note, however, that Rawls is not consistent in his formulation – see the careful analysis by Tremmel (2009: 156–7).
54. For example GiIlespie (1997: 124).
55. In Dobson (ed.) (1999: 73). Interestingly, the Stern Review on Climate Change (2007: 42) reflects closely Brown Weiss's concept of intergenerational equity. Stern (2007: 42) argues that while some resources are substitutable, at the basic level 'the global environmental and ecological system', which provides us with life 'support functions, *cannot be substituted*' (emphasis added).
56. In Hiskes (1995: 15).
57. Building on communitarian theory, Hiskes (2009) argues that a shared sense of community is linked to a nation's identity and values. He invokes a concept of 'reflexive reciprocity' (Hiskes 2009: 58–63). His argument is that by taking action now to conserve the air, water and soil quality, we will in turn protect these elements for future generations to enjoy as well (Hiskes 2009: 60). He further argues that there is a reciprocal relationship of identities involved as well in that an obligation to future generations is linked to a continuation of our own cultural identity (Hiskes 2009: 65). Hiskes' first argument when applied to climate change seems difficult to sustain. It is true that action now to curb GHG emissions will to some extent benefit those alive today as well as those living in the future. Thus for example reducing carbon emissions will immediately reduce air pollution and related health problems. But where is the reciprocity in this? If reciprocity is defined as a mutual exchange of benefits, Hiskes' argument is unconvincing. Hiskes' argument rests on a notion of reciprocity which seems too divergent from other definitions. Future generations will not convey any advantage to current generations. Moreover, deep cuts in GHG emissions taken now for the sake of those in the future will involve an imbalance in that the greater economic costs of such action will be incurred now with the greatest benefits flowing to those born in the future.
58. Referred to in Harris (2010: 103). (Attfield R, *Environmental Ethics: An Overview for the Twenty-First Century* (Polity 2003)).
59. Thank you to Robert Kirby for pointing this out. For a defence of the cosmopolitan approach, see Caney (2005).
60. The 'assets' must be conserved so that 'those coming after receive equal assets' (Brown Weiss 1989: 3, 17, 25).
61. See Gillespie (1997: 68–71).
62. Gillespie (1997: 73–6). For Islam and other religions, see Brown Weiss (1989: 18).
63. Barry takes an opposite view, arguing that sustainability is a necessary condition for intergenerational distributive justice (Barry 1999: 106).
64. See Eckersley (1998: 374) and Hay (2002: 27–9).
65. Referred to in Hay (2002: 60).
66. Hayward (2005: 35) provides the example of the 'protection of individual members of a species' being 'incompatible with protecting the flourishing of the community or ecosystem in which they live', relying on Callicott (1980). (Callicott, J, 'Animal Liberation: A Triangular Affair,' 2 *Environmental Ethics* (1980) 311). Hayward makes this argument in the context of defending a human right to a healthy environment.
67. This flows from the notion of generations being overlapping, not consecutive, discussed in 1.4.1 above. See also Wissenburg (1999) in Dobson (1999: 174).

3. Content of justice-based obligations towards future generations in the context of climate change

3.1 INTRODUCTION

I argued in the previous chapter that contemporaries have an ethical obligation to take climate change mitigation measures for the protection of future generations. This obligation rests on a harm avoidance principle, core human rights to ensure human dignity – to which persons are entitled regardless of when and where they are born – and a trans-generational community extending into the future. These broad obligations are, however, incomplete and insufficient to deal with the distributional justice issues inevitably involved in determining how the mitigation burden should be fairly distributed between current and future generations, including the rate at which mitigation should occur.[1] These distributional justice issues are the central concern of this chapter. The approach of this chapter is to identify a number of justice principles which must be met to deliver justice for future generations in relation to climate change. The justice principles provide a basis for assessing the current international climate change regime (pursued in Chapters 4 and 5). The principles also provide the basis for examining how the international regime should be reformed (Chapter 7).

The principles proposed are abstract and have – to varying degrees – found concrete reflection in the UNFCCC, including, for example, the precautionary principle and the duty of cooperation. But inclusion of principles in this manner does not guarantee that they have a deeper legitimacy, which can only be demonstrated if they reflect widely shared notions of ethics or justice. There are interesting parallels here with human rights law. Inclusion of particular rules as part of international human rights law is not enough in itself to demonstrate the legitimacy of such rules. Legitimacy requires that such rules reflect widely shared values.[2] The latter is important as it provides a basis for legitimacy, essential for such rules to ensure compliance (Franck 1995: 26). My starting point is that similar considerations apply in relation to international law rules relating to climate change.

The principles proposed in this chapter are not derived from one single theory of justice, but rather are derived from three overarching principles. These overarching principles are a *core human rights principle*, a *responsibility for harm principle* and a *capacity to pay principle*. The *core human rights principle* provides that all members of current and future generations have a human right to life, health and subsistence essential for human dignity, with a corresponding obligation on governments to safeguard these moral rights. The *responsibility for harm principle* requires one to clean up one's proportionate share of damage. The *capacity to pay principl*e provides that those with the most resources should contribute the most to addressing an environmental issue. As in Chapter 2, one may object that these theories are based on the Western philosophical tradition and therefore inadequate as a basis for establishing a global consensus on climate change mitigation. However, it was pointed out (2.1) that human rights are owed to all persons, regardless of where and when they live, and reflect widely shared values. Furthermore, the principle of 'equality' is integrally connected with human rights in the sense that all persons are *equally entitled* to core human rights. It is thus an integral component of widely shared core human rights. Thus the Universal Declaration of Human Rights refers to 'the equal and inalienable rights of all members of the human family'.[3]

The principles of *responsibility for harm* and *capacity to pay* are logical corollaries of the principle of equality. A failure to follow the *responsibility for harm* principle would mean that inequality is magnified because a person who has benefited from an activity does not take responsibility to clean up the resultant mess but rather this burden is shifted onto another person (Shue 1999 [2010]: 103; see 3.5.6 below). Similarly, a failure to follow the *capacity to pay* principle would involve magnifying inequalities as this would entail flat rates of contribution, which would result in outcomes which ignore parties' differing starting points (Shue 1999 [2010]: 106; see 3.5.8 below).

Recall that it is clear from climate change science that a total decarbonisation of the global economy by 2050 is required to avoid dangerous levels of harm. The distributional justice issue is therefore how to allocate fairly a limited 'resource' – permissible GHG emissions or, in shorthand, 1 trillion tonnes of CO_2 (Chapter 1). The issue can also be characterised as how to allocate access to the atmospheric sink, the atmosphere capacity of which is limited (3.5.4 below). The justice principles proposed apply equally regardless of whether the distributional issue is couched in terms of a limited allocation or 'budget' or limited atmospheric sink. This is because a restricted allocation of GHG emissions is required to preserve the atmospheric sink.

This chapter maintains that the three principles outlined above need to be combined with the distributional justice principles entailed in *sufficiency* and *equality* views of justice in order to apportion fairly the global GHG budget between now and 2050. Put simply, *sufficiency* views of justice put priority on achieving a basic subsistence level for as many people as possible (Page 2006: 85–95). The sufficiency notion of justice is a corollary of the core human rights to life, health and subsistence in that the latter entails the notion of a threshold below which no person should be allowed to fall. In assigning appropriate climate change burdens, a *sufficiency* principle involves ensuring that the poor – both contemporary and future – are able to achieve a basic subsistence level of existence.

Equality approaches to justice aim at equalising social economic opportunities or outcomes (Page 2006: 79–80).[4] My methodology in applying a notion of equality to the distribution of GHG emissions involves a number of assumptions. Firstly, that GHG emissions correspond roughly with economic development (3.3 below). Secondly, that the global atmosphere is a public good, appropriate for application of an equal access argument. And thirdly, the widely shared value of 'equality' in relation to core human rights provides a basis for extension of equality in relation to the distribution of GHG emissions. The latter is crucial because it underpins the legitimacy of the equality approach suggested in this chapter.

The *sufficiency* view of justice has legitimacy as an extension of the core human rights to life, health and subsistence essential for human dignity. The legitimacy of both the equality and sufficiency concepts of justice has been endorsed – in various formulations – by the global community in a number of international instruments. This includes, for example, the UN Millennium Development Goals adopted by the UN General Assembly by consensus in 2000.[5] It includes an affirmation as a 'fundamental value' (para 6) of the notion of 'equality' and that no individual and no nation must be denied the opportunity to 'benefit from development'. Chapter III of the same declaration, entitled 'Development and poverty eradication', has an implicit notion of sufficiency, in the objective of halving by 2015 'the proportion of the world's people whose income is less than one dollar a day and the proportion of people who suffer from hunger' (para 19).

This chapter argues that *sufficiency* and *equality* notions of justice provide a sound basis for distributing the mitigation burden between current and future generations. What this means is that in apportioning the global budget of GHG emissions between now and 2050, this should be done primarily on an equal per capita emission basis (3.5.9 below). However, the poor should be required to reduce emissions last, given the need to respect their sufficiency requirements tied to GHG emissions. The responsibility for harm principle, in light of the historic emissions of industrialised countries, means that these countries should bear the major share of funding costs for meeting such targets (Pickering and Barry 2012) (below 3.5.6). Both *sufficiency* and *equality* principles point in the same general direction in terms of industrialised countries bearing the lion's share of responsibility for climate change mitigation, although this will need to be modified as developing countries' emissions increase. However, the imperative of environmental effectiveness means that although current and future generations have equal value, the mitigation burden assigned to current and future generations cannot be equal: current generations must do proportionally more than their immediate forebears to avoid dangerous anthropogenic climate change.

While the focus here is on intergenerational justice, this issue inevitably becomes entangled in issues of intragenerational justice, as one argument is that the needs of the poor now should take priority over the needs of future generations. Indeed, the focus on poverty alleviation remains central in recent high-level policy and law instruments, such as the outcome of the Rio+20 Conference.[6] The justice issues addressed in this chapter relate to substantive justice. The procedural justice issues described in Chapter 1 are further addressed in Chapter 7.

This chapter is structured as follows. Section 3.2 examines key threshold issues, including the issue of who is the bearer of a justice obligation (states or individuals), and the commodity which is being made subject to distributional principles (for example GHG emissions). Section 3.4 examines the overall effectiveness requirements of a global climate regime, within which distributional justice principles must operate. It is argued that effectiveness requires duties of precaution and cooperation. In section 3.5 duties of fair distribution are examined. The *core human rights principle*, justified in section 2.1 as reflecting widely shared values, entails ascribing equal value to persons regardless of when they happen to be born (3.5.1). This principle requires for its implementation an obligation to implement sustainable development (3.5.2), a duty to ensure equality of access to the atmospheric sink (or space) (3.5.4) and an equal per capita emissions approach (3.5.9). A *responsibility for harm principle* is proposed (3.5.6) according to which a person who caused harm is responsible for clearing up the harm and/or compensation proportionate to their share of the damage. Various objections to the principle are discussed, including the claim that a lack of knowledge by industrialised countries prior to 1990 that they were harming the environment by their GHG emissions exonerates them for responsibility for their historic emissions (3.5.7(b)). Also discussed is the objection that contemporaries cannot be held responsible for their forebears' acts (3.5.7 (c)). A *capacity to pay principle* is proposed, according to which the parties with the most resources should contribute most to a common endeavour (3.5.8). Various less convincing approaches are critiqued. These include the notion of 'subsistence emissions rights', which is problematic as it assumes no upper limit to emissions (3.5.10). Also problematic is the notion of equalising the marginal costs of climate change mitigation for each nation, which is flawed as it assumes the fairness of current emissions levels (3.5.11). The discussion then turns to the extent to which a climate regime should itself seek to address issues of global poverty (3.6). In concluding (3.7), it is argued that a notion of justice that requires each generation to make equal proportionate efforts – compared with prior and future generations – in terms of climate change mitigation is untenable given the effectiveness imperative in which justice principles must operate.

3.2 WHOSE OBLIGATION?

A threshold question is who is the *bearer* of a justice obligation? The harm avoidance principle discussed in Chapter 2 placed an obligation on those exercising political power in the current generation to refrain from causing harm. Given that the focus of this book is on international law, the subjects of which are states, the key focus will be on *international justice*, that is, the level of mitigation to be borne by states. However, the ethical principles discussed in this chapter are ideally conceived of as applying at an individual level where individual human beings are the bearers of the proposed justice-based obligations, while implementation of these principles in the real world must be achieved through collective entities, including states (Page 2006: 173). This chapter is concerned with intergenerational justice defined as how the mitigation burden should be distributed by the current generation vis-à-vis future generations (Chapter 1). Our focus will be on current generations' obligations, exercising their duties through 'states'.

3.3 JUSTICE WITH RESPECT TO WHAT?

A further threshold issue is: justice with respect to what? What is the commodity which is being made subject to distributional principles? Some have suggested that distributing GHG emissions is a shorthand for distribution of the benefits or the right to engage in emission-generating activities or even economic development itself (Meyer and Roser 2010: 232). While there does seem to be a rough correlation between emissions and economic progress (Janssen et al. 1992, Tremmel and Robinson forthcoming: 19), this correlation is only very rough, as economic progress is an overly narrow measure of 'development', and energy is only one component of 'development'. There is thus no one-to-one relationship between GHG emissions and economic progress (Tremmel and Robinson forthcoming: 2). Nevertheless, it is the correlation – albeit rough – between economic progress and GHG emissions which accounts for why justice-related issues to addressing climate change rapidly merge into discourses about justice and development (3.6 below). The approach adopted here is to examine distributional issues in relation to GHG emissions themselves, on the basis of their rough correlation with economic progress. Furthermore, it is recognised that sensitivity to the broader development issues at stake is necessary in order to seek solutions considered to be fair.

In this chapter I examine distributional justice issues, firstly, in relation to a limited budget of GHG emissions and, secondly, the global atmosphere conceived as a sink which can only absorb a certain amount of GHG emissions without dangerous harm occurring to the global climatic system (3.5.4 and 3.5.9 below). In each of these framings of the issue, the justice principles will be applied directly to GHG emissions given their rough correlation with economic development.

Before discussing these justice principles, it is necessary to be clear about the overarching objective within which these principles operate, and to this issue we now turn.

3.4 ENVIRONMENTAL EFFECTIVENESS: AVOIDING DANGEROUS ANTHROPOGENIC HARM TO THE GLOBAL CLIMATE

3.4.1 Effectiveness

Effectiveness is a key requirement for delivering justice for future generations because, without an effective international climate change regime, future generations who did not create the climate change problem will suffer significant harm. Thus effectiveness is linked to the harm avoidance principle posited in Chapter 2. It is also linked to the notion of human rights; current generations have an obligation to put in place an *effective* international regime in order to protect the human rights of future generations. Thus effectiveness is central in evaluating what justice requires of current generations towards future generations in addressing climate change.

I would propose the following *effectiveness imperative* (EI), which is derived from Article 2 of the UNFCCC:[7]

EI: *An international climate regime is effective if it involves stabilization of greenhouse gas concentrations in the atmosphere at a level that would prevent dangerous anthropogenic interference with the climate system.*

Deciding what constitutes 'dangerous anthropogenic interference' necessarily involves a value choice (Oppenheimer and Petsonk 2005: 195, 205). Defining what is 'dangerous' has involved pointing to geophysical changes and biological responses, but also socio-economic and even cultural responses (Oppenheimer and Petsonk 2005: 206–8). On the face of it, pointing to the risk of *irreversible* damage has an attraction in terms of identifying a concrete threshold and meshes with the concept of tipping points referred to in Chapter 1. However, a complication here is that some climate-change-related processes – such as the destruction of coral reefs by thermal bleaching – may already be irreversible (Oppenheimer and Petsonk 2005: 195, 208). The international community agreed at the Copenhagen climate change Conference of Parties (COP) that keeping the earth's temperature below a rise of 2°C was a suitable way of defining precisely this objective (United Nations 2009, Copenhagen Accord para 2), even though the national mitigation commitments of countries announced at Copenhagen, and since, fell short of this objective.[8] As we saw in Chapter 1, this translates into a global budget of 1 trillion tonnes of CO_2 available up to 2050.

A reader might consider that this is just stating the obvious. But environmental effectiveness is crucial. This is particularly so given that one of the strong discourses in the global climate change negotiations has been a discourse of emission rights linked to a right to development (see Chapter 6). This discourse, if not placed within an effectiveness framework, carries the risk that future generations are sacrificed for the sake of contemporaries (below 3.5.10).

An effectiveness principle is also an essential part of a strong argument for rejecting high discount rates advocated by some economists (Chapter 1).

3.4.2 Duty of Precaution

The so-called 'precautionary principle' (PP) involves the notion that '[w]here there are threats of serious or irreversible damage, lack of full scientific certainty shall not be used as a reason for postponing cost-effective measures to prevent environmental degradation' (Rio Declaration on Environment and Development, Principle 15). The precautionary principle has legitimacy as an ethical obligation as it is a logical consequence of the harm avoidance and core human rights principles posited in Chapter 2. If we wait for a high level of certainty in the science, this will involve acceptance of a high risk of inflicting (avoidable) damage to the basic interests and core human rights of future generations. On the other hand, if action is taken to mitigate climate change impacts, and the climate change problem turns out to be not as harmful as expected, this would be a comparatively benign result (Shue 2010: 148).

We have seen from Chapter 1 (1.1) that there is *not* scientific uncertainty in this field in the sense that it is widely accepted by climate scientists that anthropogenic GHG emissions cause significant and long-term harm. Strictly, therefore, the precautionary principle does not apply. There is, however, uncertainty in the rate of climate change and regional impacts (1.1.3 above). This uncertainty has been used by some to create a perception of uncertainty in the broader sense and, in combating this perception, the precautionary principle retains value to ensure justice for future generations.[9]

The ethical legitimacy of the precautionary principle in the climate change context is reflected in its incorporation in the UNFCCC (Article 3(3)) as follows:

> The Parties should take precautionary measures to anticipate, prevent or minimize the causes of climate change and mitigate its adverse effects. Where there are threats of serious or irreversible damage, lack of full scientific certainty should not be used as a reason for postponing such measures, taking into account that policies and measures to deal with climate change should be cost-effective so as to ensure global benefits at the lowest possible cost.

The PP as a principle of international law is discussed in Chapter 4.

The duty to avoid harm to future generations flowing from climate change and to respect their human rights entails a duty to create the necessary institutions or reform such institutions that can deliver this obligation. This follows from application of Henry Shue's notion of 'mediating institutions' which 'stand between the right bearers and duty bearers' (Shue 1988: 703), thus providing a bridge between individual responsibilities and the establishment of necessary institutions to implement these obligations at an institutional level.[10]

While Shue's analysis was based on implementing the right to food, a similar argument can be made here based on the interdependence between individuals and institutions in relation to climate change. Thus we saw in Chapter 2 that there is arguably a transgenerational community in relation to climate change extending into the future, based on the causal relationship involved with current generations potentially harming future generations by a failure to mitigate GHG emissions. Moreover, such a community is recognised by the obligation to ensure intergenerational equity in the UNFCCC (1.10) and in the doctrine of sustainable development (3.5.2 below), even though the implementation of these principles remains very weak.

A further extension of the duty to create appropriate institutions is the duty on states to cooperate in creating an effective global regime.

Because of the global nature of the climate change problem, an effective climate change regime will only be put in place if countries are willing to make compromises which involve working in good faith towards a shared objective, where the effectiveness of the overall climate change regime is paramount (Posner and Weisbach 2010: 178). The following ethical *duty of corporation principle* (DOC) is therefore proposed as a corollary to the effectiveness principle:

> States are under an ethical obligation to cooperate in good faith in establishing an international regime which achieves the effectiveness imperative (EI).

This principle is reflected in international environmental law and the climate change regime (see Chapter 4). However, it does not in either its ethical or legal form take us very far as it begs the question as to *how* the burden of mitigation should be distributed, and it is to this distributional justice issue we now turn.

3.5 DUTIES OF FAIR DISTRIBUTION

My central argument is to propose three overarching justice principles essential for delivering intergenerational justice with respect to climate change mitigation. These are: (1) a *core human rights principle*; (2) a *responsibility for harm principle*; and (3) a *capacity to pay principle*. My argument is that an international climate change regime must combine these principles, together with equality and subsistence distributional justice principles, in order for it to meet the minimum requirements of justice.

3.5.1 Core Human Rights

The *core human rights principle* (following Caney and Shue – see Chapter 2) may be expressed as follows:

PI: *All persons born now and in the future have an equal right to life, health and subsistence.*

This principle is based on a non-discrimination principle and is linked to the notion of respect for human dignity. As Simon Caney points out, to ascribe lesser value to persons just because they are born in the future entails discrimination which is as unjustifiable as discrimination based on race or gender (Caney 2009: 168). When a person happens to be born is irrelevant to their value as a person.

The corollary to PI is an obligation on governments to ensure that persons are guaranteed rights to life, health and subsistence. However, ascribing equal value to persons regardless of when they happen to be born cannot be equated to policy-makers necessarily having to give equal weight to current and future unborn generations. To give equal weight to future generations in this manner would involve the interests of future generations swamping the interests of contemporaries (Barry 1977: 273). However, PI does entail an ethical or normative approach to 'discounting' (see Chapter 1).

Implementation of the *core human rights principle* requires adoption of three further sub-principles. These are: IP(i), an obligation to implement sustainable development; IP(iii), a duty to ensure equal access to the atmospheric sink; and IP(iv), a duty to implement a modified equal per capita emissions approach.

3.5.2 Sustainable Development

Acceptance of an ethical obligation to avoid harm in the climate change context entails an implementation principle expressed as follows:

IP(i): *All persons born now and in the future have collective rights to sustainable development. Governments have a corresponding ethical obligation to implement sustainable development.*

This principle is entailed by the harm avoidance principle, as without sustainable development future generations' basic interests and core human rights would be harmed. Crucial in this sub-principle is the definition of sustainable development. There is a bewildering array of definitions. However, the most widely recognised definition of sustainable development is that of Brundtland as: 'development that meets the needs of the present generations without compromising the ability of future generations to meet their own needs' (World Commission on Environment and Development 1987: 43). This definition entails a notion of intergenerational justice.

The Brundtland definition, however, leaves unclear what precisely is to be sustained, and whether the obligation is to maintain overall human well-being or merely to ensure that one's successors have the ability to lead reasonable lives above a basic subsistence level.[11] Given the range of definitions of sustainable development, the approach taken here is, following Carmody (2012: 70), to take a broad definition of sustainability which encompasses many narrower definitions. Sustainable development is defined as embracing: 'economic and financial stocks'; 'human stocks' (including 'wealth embodied in labour and skills'); 'environmental stocks' (which include ecosystems and the atmosphere); and 'social stocks' (which include 'institutional arrangements' and 'interpersonal networks, individual rights and freedoms') (Carmody 2012: 70–71).[12]

Sustainability involves the 'maintaining or increasing' of well-being across generations, including in relation to these elements. In order to achieve this, trade-offs may be necessary for example between or within environmental and economic stocks. This has led to two different versions of sustainability – 'strong' and 'weak' sustainability. 'Weak sustainability' is where all the different forms of stocks are considered to be completely substitutable. In contrast, 'strong sustainability' is where no substitution is permitted between these various stocks (Carmody 2012: 72). Real-world policy-making tends to lie 'somewhere between these two extremes' (Carmody 2012: 72).

Strong sustainability should apply at least in relation to stocks, such as elements of the ecosystem upon which human beings depend, that are not substitutable (Carmody 2012: 73), as otherwise there is a risk of well-being suffering. However, it can be difficult to determine precisely which elements are substitutable.[13] Avoiding dangerous anthropocentric climate change is required by strong sustainability given human beings' well-being is dependent on ecological systems of which the global climate system is a vital part.[14] The UNFCCC states that parties 'have a right to, and should promote sustainable development'. What this means as an international law obligation is discussed in Chapter 4.

We have seen that sustainable development has embedded in it an ethical obligation on current generations in relation to future generations. However, it is important to define more clearly exactly how the mitigation burden should be distributed between current and future generations in relation to climate change, and to this issue we now turn.

3.5.3 Duty to Balance Current and Future Generations' Interests

While sustainable development, as defined above, entails an obligation towards future generations, it remains unclear what weight should be given to the respective interests of current and future generations in climate change mitigation. This issue is difficult given that a shift to a carbon-neutral economy cannot be achieved without at least short-term serious social dislocation and suffering. Identifying justice principles that can resolve potential conflicts between current and future generations in this context has been relatively unexplored (Gardiner 2004 [2010]: 15).

Some broad principles seem clear. Firstly, so-called 'no regrets' climate change action should be taken immediately by governments, in other words measures with short- and long-term economic and mitigation benefits such as energy-saving measures (Gardiner 2004 [2010: 13). Secondly, governments need to work towards restructuring their economies so that a rapid move to a zero-carbon economy becomes possible within the time frames indicated by scientists. How these policy imperatives should be reflected in an international climate treaty regime is a theme taken up in Chapter 7.

In relation to weighing the interests of future and current generations' interests in relation to climate change, I would propose the following *structural reform implementation principle*:

IP(ii): *Policy-makers should not delay structural economic reforms necessary to ensure the protection of the long-term interests of future generations while minimising harm to the core rights of current generations.*

The idea behind this principle is to place an onus on policy-makers to implement sustainable development in a manner which minimises harm to the core rights of current generations. By 'core rights' I am referring to the rights to life, health and subsistence (see Chapter 2).

The structural reform principle rests on a utilitarian argument combined with the *core human rights* principle expressed above which entailed persons of current and future generations having equal value. The latter notion of equality can be expressed within an individual human rights framework (Caney 2009 [2010]: 163) or in a utilitarian framework. Tremmel's theory follows the latter in his principle of maximising 'the average wellbeing of all members of all generations' (see Chapter 2). A weakness in the latter is that it could be used to justify sacrificing particular social groups as long as the overall average well-being increases. For example, a utilitarian approach could be used to support the argument that current generations' well-being should be entirely sacrificed for future generations if this delivers increased overall well-being for human beings as the *number* of people born in the future will exceed that of current generations (Barry 1977: 273).

Such a utilitarian approach would conflict with an individual human rights approach whereby individual rights are inviolable and cannot be traded off against broader social goals.[15] Moreover, a utilitarian approach of this nature would violate the principle of equality of core rights across generations posited above as it would imply priority being given to future generations over and above current generations.

If sustainable development is implemented with a shift to renewable forms of energy, the potential conflict between current and future generations will lessen. However, the longer there is a delay in making the required deep cuts in GHG emissions, the greater the chance of difficult choices having to be made between current and future generations' interests (Shue 2011: 313).

3.5.4 Duty to Ensure Equality of Access to the Atmospheric Sink

The core human rights principle posited above entails an obligation to ensure equal access to the atmospheric sink or space.[16] Thus, Neumayer has argued for a principle which may be summarised as follows:

IP(iii): *All persons have an equal opportunity to use the absorptive capacity of the planet 'no matter where or when he or she happens to live' (Neumayer 2000: 188).*

Neumayer's argument rests on science to the extent that climate change is a product of cumulative emissions over a long period of time (see Chapter 1). He notes that there are limits to the natural absorptive capacity of the earth in terms of the gradual decay of certain GHGs over time.[17] This absorptive capacity of the earth belongs to nobody and should 'therefore be equally assigned to everybody in order to give everybody equal opportunity to benefit from emissions' (Neumayer 2000: 188).

Singer argues in a similar manner, using the analogy of a giant sink in a village which is used for waste (Singer 2002 [2010]: 187). Initially waste material is put into the well by villagers without any apparent negative impacts. Some villagers are big consumers and produce a lot of waste; others produce very little. After a certain time it becomes clear that the waste has been accumulating and it begins to cause toxic impacts on a waterhole where children swim and it becomes clear that the village water supplies are threatened (Singer 2002 [2010]: 187). The 'equality of access' principle means that all of the villagers have equal access to the public resource of the sink. However, if a minority of the villagers with high consumption have used up 75 per cent of the sink capacity, *fairness* dictates that this minority of the villagers should be primarily responsible for addressing the problem as they received most of the benefits of the sink and used more than their share: assuming each villager has the same entitlement to the sink (Singer 2002 [2010]: 188).

Applied to climate change, the 'equal share of the atmosphere' argument translates into industrialised countries having the primary responsibility to address climate change as they have exceeded their share of the global atmospheric resource (Singer [2010]: 189). Singer relies on John Locke, who argued in 1690 that the earth and its resources were common property; appropriation of this common property could occur provided that there was 'enough and as good left in common for others' (Locke 1690 [1980]: 19).[18]

What are the intergenerational justice implications for this principle of equality of access? The principle IP(ii) applies regardless of spatial or temporal constraints 'no matter *where* or *when*' (my emphasis) a person happens to live. To ensure that persons who comprise future generations have an opportunity to use the absorptive capacity of the planet *equal* to persons currently alive, strong mitigation action must be taken consistent with the *effectiveness imperative* (EI) set out above. The reason for this is that without such action future generations will be denied such equal opportunity. Moreover, equal access to the absorptive capacity of the atmosphere is also a precondition for the enjoyment of future generations' core human rights. Conversely, a failure to mitigate climate change carries the risk of an eventual *disruption* of the absorptive capacity of the planet, carrying with it a failure of current generations to meet the duty to protect the human rights of future generations.

Into the future, each generation which has exceeded use of their fair share of absorptive capacity – and benefited therefrom – will have a heavier burden than others to take mitigation action. Thus over time, the burden will gradually shift from industrialised to developing countries.

3.5.5 Objection: Assumed Equality of Needs

Various objections have been raised in response to the 'equality of access to the environmental sink' and 'responsibility for harm'/historic accountability arguments. One objection to the former is that people's *need* for access to the atmospheric sink cannot be assumed to be the same. Thus, for example, people in colder climates have a higher need for higher GHG emissions related to heating than those in warmer regions (Vanderheiden 2008: 108). One response to this is that these factors cancel each other out – for example the need for heating in some regions is balanced by the need for air-conditioning in hotter parts. A further argument is that this approach can be the basis for appropriate national emission targets which can then be modified to take into account particular geographical or other needs. A good analogy here is the use of rationing during the Second World War. The general rule was that each person received the same number of rations based on an assumption of roughly equal needs. However, rations were issued to those with particularly acute needs (Zweiniger-Bargielowska 2000: 15).[19] There is no entirely satisfactory response to the equality of needs problem other than to say that the 'equality of access' principle is better than alternatives which tend to be more complex to translate into obligations in a climate treaty.

A further variant of this objection is to argue that the principle of equality of opportunity ignores a right of industrialised countries to continue their high levels of development into the future similar to a 'squatter's rights' derived from the doctrine of 'adverse possession' in common law (Young and Wolff 1991, quoted in Neumeyer 2000: 188, Bovens 2011). However, this ignores the well-established doctrine that appropriation of property rights will only be regarded as in accordance with the dictates of justice if the appropriation does not entail worsening the situation of others (Neumeyer 2000: 188). In relation to climate change, the high consumption of industrialised countries has lessened the opportunities of developing countries in terms of their access to the absorptive capacity of the global atmosphere (Singer 2010: 188).

Conceiving use of the atmospheric sink as a property right can, in my view, only be on the basis of an analogy. Conventional notions of property rights include the idea that the property right-holder can exclude third parties. This clearly does not apply to the 'expropriation' of the atmosphere by pollution of GHG emissions. The atmosphere is one of the few examples of a pure public good where no one has a better claim over this natural resource than any other (Vanderheiden 2008: 225). This is not to deny that a property rights analysis is not valuable. The analogy with property rights is powerful, particularly given that use of the atmospheric resource by emitting GHGs lessens the value of the remaining resource for others (Moellendorf 2011: 107).

3.5.6 Responsibility for Harm/Historic Emissions

A closely related argument to that presented above is the claim that the industrialised countries are chiefly responsible for climate change mitigation – and therefore delivering justice to future generations – because they *caused* the problem. This argument rests on the science – CO_2 persists in the atmosphere over long periods (Chapter 1). Thus the climate change problem would not have occurred but for the *cumulative* emissions since the Industrial Revolution of industrialised countries, which account for the bulk of current GHG emissions in the atmosphere.

The historic emissions argument stated above can be formulated as part of a broader ethical *responsibility for harm principle*:

PII: *A person who caused harm to the interests of others – and no one else – is responsible for clearing up this harm and/or compensation proportionate to their share of the damage.*

The ethical basis of this principle rests on the notion that if one creates a mess and does not clear it up, this imposes a burden on another person who has not in any way benefited from the mess (Shue 1999 [2010]: 103). This is unfair to the person who does end up paying the costs of the mess and who presumably has not consented to this occurring. Thus we can see a link to equality, as the person who does clear up the mess has been put at a disadvantage for no reason. This notion is therefore a corollary of the principle of equal entitlement to core human rights (Shue 1999 [2010]: 103).

This principle is to be distinguished from the 'polluter pays principle', an *economic policy principle* which was embraced by the OECD (1974) to the effect that future costs of pollution should be internalised in prices. This principle is only forward-looking and excludes compensation (Shue 1999 [2010]: 103). The principle is also linked to utilitarianism in that, by the polluter rather than victim bearing the costs of pollution, an incentive is provided which encourages potential polluters to prevent pollution for the general good (Singer 2002 [2010]: 194).[20]

The polluter pays principle is strongly linked to the principles I have argued are essential for ensuring intergenerational justice in relation to climate change. A failure to strongly mitigate climate change by contemporary generations involves a refusal to respect the polluter pays principle with the result that future generations must unfairly bear the costs in relation to actions of contemporary generations.[21] This in turn entails future generations' human rights being impacted upon being born. Using Shue's language, it involves future generations being 'taken advantage of' by current generations, thus breaching the equality principle posited above.

Could future generations be compensated for a failure by current generations to strongly mitigate climate change and in this way the polluter pays principle be respected? The difficulty with this is that climate change involves a range of impacts which are to a large extent unable to be compensated (Chapter 1).

The *responsibility for harm principle* provides a solid basis for modifying an equal per capita emissions approach in a way whereby industrialised countries have a greater share of the mitigation burden owing to their greater contribution to the problem (see Chapter 7). Moreover, this principle can also be legitimately taken into account in compensating future adaptation costs of climate change, an issue outside the scope of this book.

3.5.7 Objections

(a) Proportionate harm

An objection to the historic accountability argument is that even if emissions from industrialised countries had stopped entirely in the mid-1980s when the climate change science became more fully known, we would still have a problem owing to the continuing emissions of developing countries (Miller 2008: 154). In my view this does not undermine the 'responsibility for harm' principle argued for above, rather it suggests that responsibility should be *proportionate* to the harm caused (see Vanderheiden 2008: 192). Thus industrialised countries remain responsible for the lion's share of mitigation, but their proportionate share of responsibility will change over time as developing countries increase their share of cumulative emissions.

(b) Lack of knowledge

Is the historic accountability argument undermined by the fact that prior to the mid-1980s, or at the latest 1990, it was not widely known that GHG emissions caused harm? Some have argued that there can be no ethical responsibility without knowledge that foreseeable harm was being caused (Caney 2005 [2010]: 130–31). Thus, for example, Tremmel argues that '[f]ault implies intent or negligence, and neither is present' (Tremmel and Robinson forthcoming: 18).[22] This is, however, unconvincing. In national legal systems, responsibility for various types of environmental harm is imposed where the harm has been caused without knowledge or intent, for example, in so-called strict liability environment-related offences (Baer 2006 [2010]: 250). Miller describes this type of responsibility on an ethical plane as 'outcome responsibility' (Miller 1999: chapter 4).

Shue (1999 [2010]: 104) argues that the objection of ignorance merges together punishment for an action and being held responsible for an action. While it would indeed be unfair to punish someone for actions they could not have known were harmful to others, it is not unfair to make them pay the costs given they caused the problem. In response, Caney questions whether, if one should not punish persons causing harm, 'why is it alright to impose financial burdens on them?' (Caney 2005 [2010]: 131). But this overlooks the fact that national legal systems do distinguish between these types of responsibility *albeit* with punishment possible in relation to particular offences even absent an intention to cause harm. A good example is the offence of 'manslaughter', where a person causes the death of another without intention, which is subject to punishment, albeit with a lesser sentence compared with the offence of murder, where there was an intention to kill.[23] Thus ignorance may justify compensation short of punitive damages which respects the lack of knowledge by the person creating the harm, but also respects the need to compensate the victim.

(c) Making the current generation responsible for their forebears' acts

A further objection to historic accountability is that it involves imposing ethical obligations on persons for acts done by their predecessors. How can it be fair for a person alive in the US today to be made responsible for GHG emissions related to activities of their forebears which occurred before they were alive?

At an *individual* level there does seem force in this argument. For example, children are not made liable for the debts of their parents or grandparents. However, at a collective level the situation is arguably different. In terms of collective responsibility we do ascribe responsibility for states spanning across generations. Examples can be found in relation to war reparations and apologies by governments for previous injustices (Meyer 2008). A further example is intergenerational transmission of debt; sovereign debt can remain owed by a country over successive generations (Pickering and Barry 2012). An example of apologies by governments for previous injustices is the Australian government's apology in 2008 relating to stolen Aboriginal children forcibly removed from their parents.[24] This apology can only make sense on the basis of some collective notion of responsibility. Moreover, it can be argued that while there have been great changes in relation to some states which challenge the notion of a continuing moral agent (for example breaking up of states in Eastern Europe, mass migration to South America, and so on), there is still considerable continuity in the sense that in Western Europe, Japan, the US, Canada and Australia, those born into these societies generally enjoy a range of benefits not enjoyed by many in developing countries (Neumayer 2000: 189).[25]

Moreover, notions of *collective* ethical responsibility for harm are embedded in international law – a good example of this is Principle 21 of the Stockholm Declaration, which makes states responsible for environmental harm caused to areas outside their jurisdiction.[26] The ethical notion underlying Principle 21 enjoys strong support and Principle 21 is now part of customary international law (Sands and Peel 2012: 197), although its application to climate change is problematic for various reasons (see Chapter 4).

The historic responsibility argument that current members of industrialised countries should bear the burden of climate change mitigation is linked to the idea that these countries have created an inequality by imposing costs on other people, which can in turn justify these countries taking on an additional burden to *correct* the inequality (Shue 1999 [2010]: 104). A related argument is supported by the notion of equality of opportunity, which as we saw above is used to support the argument that industrialised countries have used up their equitable share of the global absorptive capacity of the atmosphere and that from this flows their responsibility to act first to reduce emissions, 'while allowing developing countries to achieve at least a basic level of development' (Agarwal et al. 2002: 173).[27] Caney argues that these arguments both rely on a collectivist approach, which is not warranted. In support of taking an *individualist* approach he gives the following story. Consider two families, each of which has a son. Imagine that several generations ago one of the families sent their child to a prestigious school (for example Eton) whereas the other family sent their child to an ordinary school. Following the individualist approach, he states that:

> the fact that someone's great-great-great-grandfather enjoyed more than fair opportunities does not give us any reason to give them a less than equal opportunity. But the collectivist position is committed to claiming that we should penalize the descendant. It must say that since one family had a greater than fair allocation of educational opportunities in the past, this must be rectified by giving it (or, rather, one of its members) a less than equal opportunity now. But that seems just bizarre and unfair (Caney 2005 [2010]: 133).

This analogy is not convincing as – unlike the global absorptive capacity of the earth – access to education is not an exhaustible resource diminishing over time. The sending of a child to a prestigious school by previous generations does not *causally* diminish the total pool of places available at schools now.[28] On the other hand, emission of GHGs by previous generations in industrialised countries has had the causal impact of lessening the opportunities for current generations, particularly in developing countries, to develop now. It is this interdependence which gives force to the argument that – at a collective level – the countries that have exceeded their share of the common resource should first make amends to redress the inequality.

(d) Benefits of emissions

A further objection to the polluter pays principle referred to above is the argument that developing countries have also benefited from the emissions of industrialised countries. Technological inventions have flowed from industrialised and developing countries or, put differently, 'the benefits of emissions have spread beyond the emitters' (Grubb 1995: 491). However, in general terms, most of the benefits of high GHG emissions in industrialised countries have flowed to the industrialised countries themselves (Singer 2002 [2010]: 192). It would seem reasonable that, at most, some modest discounting should occur to take into account this flow of benefits to developing countries (Page 2006: 169). Moreover, the harm caused by emissions outweighs any associated benefits.

For those who reject the historic emissions argument, a 'capacity to pay' argument is often invoked as a basis attributing ethical responsibility for industrialised countries to bear the main mitigation burden in relation to climate change, and to this we now turn.

3.5.8 Capacity to Pay

Shue proposes the following principle:

PIII: *Among a number of parties, all of whom are bound to contribute to some common endeavour, the parties who have the most resources normally should contribute the most to the endeavour (Shue 1999 [2010]: 105).*

Shue acknowledges that it is probably not possible to derive this principle of capacity to pay from even more fundamental principles and it will be observed that the principle already includes within itself the assumption that the parties in question are 'bound to contribute' (Shue 1999 [2010]: 105). Nevertheless, he points out the link between this principle and the principle of equality. Shue demonstrates that the alternative to 'capacity to pay' would be equal or flat rates of contribution to a particular endeavour which – by disregarding the starting position of each person – would lead to a worsening of inequality in outcomes (Shue 1999 [2010]: 106). Thus we can see that the 'capacity to pay' principle logically flows from the premise of all persons being equally entitled to core human rights essential for human dignity.

Page points out that it is difficult for the 'capacity to pay' principle on its own to support an obligation on industrialised countries to pay the costs of climate mitigation. He notes that if one imagines that climate change impacts were real but *without* anthropogenic sources, the developed world, according to this principle, would be equally responsible (Page 2006: 172). This seems counterintuitive as it does seem to make a difference that the industrialised countries have caused the problem. Page sensibly concludes that a pluralistic approach giving weight to both historic responsibility and capacity to pay is most convincing (Page 2006: 173).

The capacity to pay claim is often combined with a subsistence argument. Inhabitants of industrialised countries have generally more than enough resources required to lead a decent life and are therefore under a duty to assist others as they can do this without reducing their standard of living below what is required to lead a decent life (Page 2006: 171). Implicit in this argument is that the costs of addressing climate change are not so great that they would reduce well-being below this threshold.[29]

An apparent weakness of the polluter pays/beneficiary principle is that it cannot provide a basis for mitigation obligations where polluting countries fail to meet their obligations (Caney 2005 [2010]: 134). While the industrialised countries have generally failed to meet their obligation set out in the UNFCCC of taking the lead in reducing GHG emissions, as argued above, a notion of responsibility *proportionate* to the harm occurring would still entail some level of responsibility on developing countries in this context. Moreover, as some developing countries become wealthier, they would then also have a proportionate 'capacity to pay'. As mentioned (above 1.8), some have argued for a delay in taking mitigation action on the basis that future generations will be wealthier than current generations. This is in effect a version of the capacity to pay argument. This argument, however, is unconvincing. Firstly, any delay in emissions violates the effectiveness imperative (EI) (Chapter 3). Secondly, the assumption of increasing economic growth is questionable, given the potential impact of climate change and the decline in productivity growth in developed economies since the 1970s (Carmody 2012: 69).

3.5.9 Equal Per Capita Emissions

One of the most strongly supported proposals is to cap emissions at an overall level and then divide the emissions within this cap on a per capita basis, with national emission quotas based on these per capita levels (Gardiner 2004 [2010]: 16).[30] Combined with the budget approach, this 'modified equal per capita emissions' principle may be expressed as follows:

IP(iv): *Under an emissions budget necessary to meet the effectiveness imperative (EI), national mitigation quotas should be based on per capita levels, with some modification to take into account particular geographic or other needs.*

Under the 'contraction and convergence' version of this approach, emissions of all countries should over time converge at equal per capita amounts and then be reduced.[31] This approach can be combined with a budget approach within a cap which requires zero emissions by 2050.

This approach has considerable attraction in giving equal weight to all people regardless of where and when they were born and in this sense flows logically from the core human rights principle discussed in Chapter 3. However, the starting point for equal per capita schemes is an important variable. Moreover, the issue of the respective weight which should be given to current and future generations is still involved in deciding *how quickly* the emissions should converge and be reduced.[32]

The equal per capita approach is attractive in its simplicity but nevertheless has a number of important shortcomings. Firstly, it ignores geographical and other conditions which may impact on the need for energy – and GHG emissions – in different places. Societies vary in terms of their opportunity to use non-carbon energy sources (Caney 2011). The same arguments made above in relation to equal access to the global atmospheric sink apply here (Chapter 3). However, in response to these concerns, an equal per capita approach can be the basis for appropriate national emission targets which can then be modified to take into account particular geographical or other needs.

A further criticism of the equal per capita approach is that it would encourage population growth. This difficulty can be overcome, however, by simply agreeing on a reference base year. Moellendorf, for example, proposes using a projection of the global population in 2050 against a year 2000 baseline (Moellendorf 2009a: 125).

A more fundamental criticism of the equal per capita approach is that it ignores the key difference between so-called 'survival' and 'luxury' emissions. The idea here is that some GHG emissions are linked to people's very survival, whereas other emissions are used to produce luxury items (Shue 1993 [2010]: 200–214).

3.5.10 Subsistence Emission Rights

Giving priority to subsistence needs is at the heart of proposals, according to which, people should have inalienable rights to a minimum level of emissions necessary for their survival or to some minimum quality of life (Shue 1993 [2010]: 202). This suggests that there may be 'moral constraints on the limitation of emissions', with some emissions 'deemed morally essential' with the implication that these emissions need to be guaranteed even if this leads to overshooting emission targets prescribed by scientists (Gardiner 2004 [2010]: 16).

Traxler indeed argues that subsistence emissions, even if they cause harm, can be morally excusable because 'they present their potential emitters with such a hard choice between avoiding a harm today or avoiding a harm in the future' (Traxler 2002: 107–8). Traxler relies on the notion of 'social necessity', which he defines as 'what a society needs or finds indispensable in order to survive' (Traxler 2002: 106). He goes on to argue that 'where the present harm from not emitting is conspicuous enough', it would be 'unrealistic, unreasonable, and maybe even irrational to expect present people to allow present harm and suffering to visit them or their kith and kin in order that they might avoid harm to future people far less closely related to them' (Traxler 2002: 107). He likens this situation to a plea of self-defence to murder, arguing that the social necessity for subsistence excuses these emissions and the resulting future infliction of harm they cause (Traxler 2002: 107). There is clearly an acute moral dilemma here which can only be escaped if there are dramatic emissions reductions accompanied by a shift to renewable energy sources. While Traxler's approach has great intuitive attraction, it implies that current generations can trump the interests of future generations violating the effectiveness imperative (EI) posited above. Perhaps this is inevitable and underscores the urgency of action on climate change, particularly by the higher emitters.

Moreover, the very notion of a 'right to GHG emissions' is conceptually problematic as it seems to confuse means and ends. The use of energy – rather than energy produced *necessarily* with GHG emissions –

is a means of ensuring that people's subsistence rights are met. The notion of 'GHG emission rights' turns this on its head. As Hayward points out, while it makes sense to argue for a human right to a minimum level of subsistence and a corresponding obligation on governments to ensure that this is delivered, emission rights may only 'be contingently suitable means for promoting those rights' (Hayward 2007: 441). They are contingent in the sense that subsistence may be delivered increasingly by energy sources not reliant on CO_2. Hayward argues convincingly that a human right to minimum emissions would 'exacerbate rather than resolve the problem of excessive global emissions' as it implies there is no upper limit (Hayward 2007: 441). This is not to deny that human rights do not have a valuable role as a basis for justifying action to mitigate climate change.

Of course it is possible to support an approach whereby emissions are to be reduced in a manner according to which subsistence GHG emissions of the poor are given priority without enshrining this priority principle in rights language. Difficulties will remain in reaching agreement on an appropriate 'level of subsistence' which to a considerable degree may be dependent on the context of the particular society in question (Gardiner 2004 [2010]: 17). Nevertheless, an approach which requires the wealthy to take action to mitigate climate change, while the poor only take action last, resonates strongly with subsistence notions of justice which, as we have seen, are supported as widely shared values (Chapter 3) and seem most likely to gain political traction (see Page 2006: 90–95). This 'subsistence' principle can be expressed as follows:

IP(v): *In assigning mitigation burdens under an emissions budget necessary to meet the effectiveness imperative (EI), the wealthy should take action first, with those under a basic level of subsistence being required to take action last.*

3.5.11 Equal Sacrifice

A further proposal is based on equalising the marginal costs for each nation required to address climate change (Traxler 2002). Traxler argues that a key attraction of this proposal is that it places 'the most moral pressure possible on each nation to do its part' (Traxler 2002: 102). However, it is unclear whether the proposal would meet this objective, particularly given that under this proposal, previous emissions of the rich countries are ignored and it is these emissions which in the eyes of many developing countries potentially deny them the opportunity for development (Gardiner 2004 [2010]: 18). As we will see in Chapter 6, arguments based on historic emissions are very prominent in developing countries' discourses in the UN negotiations. Therefore a proposal which completely disregards this discourse seems unlikely to find widespread support.

A more basic flaw in this approach is that it assumes the fairness of the original benchmark for reductions (Moellendorf 2009a: 120). Given the failure by most industrialised countries to meet their commitment to take the lead in reducing emissions since 1992, it is difficult to see how these countries' current emissions can be a legitimate starting point for emission reductions.

A further difficulty with the equal sacrifice or 'fair chore division' approach is that it would require making assessments as to projected future economic growth without mitigation action, which remains speculative and difficult.[33]

Moellendorf argues, based on a Rawlsian approach, for an intergenerational equal burdens principle according to which mitigation action to address climate change by the current generation should only be proportionate to the costs of mitigation as a percentage of GDP which future generations will be subject to (Moellendorf 2009c). As explained in the previous chapter, it is unrealistic to expect to find a principle relating to climate change mitigation which can apply to any generation regardless of the context (Chapter 2). While initially appealing, the principle argued for by Moellendorf is unpersuasive, as the environmental imperative of avoiding dangerous anthropocentric climate change dictates that the current generation has an ethical obligation which is *proportionally greater* than the burden of previous generations. This unfairness results from the previous generation not undertaking their fair share of emission reductions. This unfairness would be compounded if the current generation undertook less mitigation than what is required from an environmental effectiveness point of view.

3.6 JUSTICE, POVERTY AND THE GLOBAL CLIMATE REGIME

Posner and Weisbach argue that the global climate regime should not seek to redistribute wealth, from the rich to the poor, but rather should only seek to solve the climate change problem (Posner and Weisbach 2010: 6). There is some validity to this view in that, at the international level, there has never been agreement by industrialised countries to undertake as an international obligation the redistribution of wealth to developing countries (Posner and Weisbach 2010: 132). It is therefore unlikely that this could be achieved through a global climate change regime.

However, it is equally true that developing countries are unlikely to agree to a global climate change regime which they perceive as unfair. This sense of fairness is linked to the polluter pays principle – the notion that industrialised countries created the problem and now have prime responsibility for solving it. Moreover, developing countries feel they have been unfairly treated by the international system in relation to a range of non-environmental issues, including debt and trade issues (Roberts and Parks 2007), and that these injustices should be compensated for in the climate change regime. While it is unclear why such injustices should be dealt with through the climate regime rather than the debt and trade regime, industrialised countries need to build trust by meeting their commitments to take the lead in reducing GHG emissions.

While a global climate change regime should not seek per se to resolve global poverty issues, it must nevertheless take into account the different situations of respective members of the global community in terms of their different levels of development. As stated earlier, GHG emissions do roughly correspond to economic development and with this goes responsibility for causing the problem and also a capacity to address it. Without being sensitive to the justice discourse, developing countries will remain unlikely to play their essential part in constructing an effective regime. This has begun to be recognised by industrialised countries, who have agreed to a USD100 billion (per year by 2020) to assist developing countries to pay for climate adaptation, some of which will flow through a green fund (Chapter 4).

3.7 CONCLUSION

This chapter has argued for a number of principles which operate as criteria for assessing whether an international climate change regime meets the requirements of intergenerational justice. The key principles of *core human rights* essential for human dignity, *responsibility for harm* and *capacity to pay* were argued to be central. These principles need to operate in conjunction with the core distributional justice principles of equality and subsistence. All of these principles rest directly – or indirectly – on widely shared values (2.1 and 3.1). The principles of *responsibility for harm* and *capacity to pay* are logical corollaries of the widely shared value of 'equality', which is an integral component of the notion of core human rights as each person is considered to be *equally* entitled to human rights (Shue 1999 [2010]). These principles in turn require a number of implementation principles, which include principles of sustainable development, equal access to the atmospheric sink and an equal per capita approach (see Figure 3.1).

Core Principles: Equality and Subsistence
Key Principles: PI Core human rights PII Responsibility for harm PIII Capacity to pay
Implementation Principles (IPs): IP(i) Sustainable development IP(ii) Structural reform principle IP(iii) Equal access to the atmospheric sink IP(iv) Modified equal per capita emissions IP(v) Subsistence principle

Note: Procedural justice is not dealt with in this figure.

Figure 3.1 Justice principles for ensuring substantive justice for future generations in relation to climate change mitigation

A difficulty in the approach suggested here is that the responsibility for harm principle would seem to push in a different direction from the more forward-looking equal access to the atmospheric sink or equal per capita approaches, when applied within a fixed GHG budget up to 2050. A response to this dilemma is to retain the notion of allocating prospective emission entitlements on an equal per capita basis but to take into account historic emissions by industrialised countries by these countries being responsible for mobilising finance to assist developing countries in both mitigation and adaptation efforts (Pickering et al. 2012: 437).

The principles proposed here require urgent and deep cuts in GHG emissions in order to deliver justice for future generations. The lion's share of the mitigation burden is on industrialised countries. However, given that over time the proportionate responsibility for emissions will in the coming decades shift from industrialised countries in the direction of developing countries, there is a need for principles that retain validity over time. The *responsibility for harm principle* meets this requirement as it is flexible enough to take into account the proportionate contributions of countries to the problem. In addition, both subsistence and equality notions of justice entail sufficient flexibility to apply into the future.

While a rapid shift to genuinely sustainable development would ensure that conflicts between the interests of current and future generations in relation to climate change are avoided, such a shift looks to be unlikely at present. The longer there is delay in serious mitigation efforts, the more acute will be the potential conflicts between current and future generations. Indeed, Gardiner has spoken of a potential 'arms race' between generations in relation to climate change (Gardiner 2011: 185–209).

An equality of mitigation burden, when applied across generations, could suggest that each generation should contribute the same amount in relation to the overall mitigation effort. However, on closer analysis this approach was found to be wanting. The effectiveness imperative (EI), combined with the requirement to avoid harm, entails that the current generation do what is required to avoid dangerous anthropogenic climate change, even if this requires doing more than previous or future generations. Given that current mitigation efforts are so seriously lacking when compared with what scientists say is necessary, there is little risk in the current generation doing *more* than required. If the current generation does take the necessary mitigation action required, this would mean a reduction in overall economic wealth transmitted to the next generation. However, this reduction would meet fairness requirements in that the reduced GDP received by the next generation would be offset by receiving a climate system that was not irreversibly damaged. Moreover, the current generation's actions – should they address climate change effectively – need to be considered in a context where the immediately prior generations have done less than what is required.

NOTES

1. Equally important justice issues relating to adaptation are not addressed here. See Baer (2006 [2010]: 247–62).
2. Besson (2012: 215) uses the term 'universal values' but I prefer the term 'widely shared values' as this avoids any suggestion of linkage to notions of natural rights or human rights being somehow integral to human nature (see discussion at 2.1).
3. Universal Declaration on Human Rights, 1948, http://www.un.org/en/documents/udhr/index.shtml, accessed 17 April 2013, Preamble para 1.
4. Another important notion of justice is the so-called *priority view,* according to which relative wealth does not matter; rather, the position of the worst off is crucial (Page 2006: 83). A serious problem with this view is that it fails to take into account the *relative* prosperity of people. Moreover, it creates obligations even where people live above a sufficiency level (Page 2006: 90).
5. Resolution 55/2 of the UN General Assembly, United Nations Millennium Declaration (UN A/res/55/2 2000), http://www.un.org/millennium/declaration/ares552e.htm (accessed 17 October 2012). The Millennium Goals were reaffirmed by the General Assembly in 2010. See UNGA Res. 65/1 (19 October 2010, UN Doc. A/RES/65/1).
6. Report on the United Nations Conference on Sustainable Development, Rio de Janeiro, Brazil 20–22 June 2012. http://www.uncsd2012.org/content/documents/814UNCSD%20REPORT%20final%20revs.pdf> (accessed 27 August 2013).
7. Article 2 of the UNFCCC provides that the objective of the Convention is: 'stabilization of GHG concentrations in the atmosphere at a level that would prevent

dangerous anthropogenic interference with the climate system. Such a level should be achieved within a time frame sufficient to allow ecosystems to adapt naturally to climate change, to ensure that food production is not threatened and to enable economic development to proceed in a sustainable manner.'

8. See Levin and Bradley (2010). Light and Pool (2010) conclude that the pledges made at Copenhagen would fail to hold global warming below 2°C but would hold warming below 3°C cf 4.8°C by 2100 with a business-as-usual scenario.
9. Shue argues convincingly for a version of the precautionary principle which relies on three particular features of the climate change problem (Shue 2010: 148). He argues that questions of probability beyond a certain minimal level of likelihood should be ignored in cases where: (a) 'the magnitude of the possible losses is massive', even though the precise probability cannot be specified, (b) 'the mechanism by which the losses would occur is well understood', (c) 'the conditions for the functioning of the mechanism are accumulating', and (d) the costs of prevention are not excessive in comparison with the magnitude of possible losses and other important claims on our resources. Shue argues that the climate change problem falls within these conditions and thus requires 'urgent' and 'effective' action to make the outcome 'progressively more unlikely' (ibid.: 148).
10. Shue's analysis is based on obligations relating to the universal right to food, where economic interconnection between persons justifies the argument based on transforming individual obligation into the required institution (Shue 1988: 702).
11. Page argues that it entails the latter (Page 2006: 91).
12. In relation to social stocks, Carmody relies on Parkinson M. (2011), *Sustainable Wellbeing – and Economic Future for Australia*, Shann Memorial Lecture Series, 23 August. Note that Carmody refers to 'sustainability' and that I use her definition to define 'sustainable development'.
13. Carmody points out it can be very difficult to determine whether a particular stock or parts of a stock are substitutable given the lack of certainty of future generations' preferences and technological developments. However, 'current experience regarding fundamental determinants of wellbeing can serve as a guide to stocks that are likely to not be substitutable below a certain point – for example basic freedom and education, supplies of breathable air, drinkable water and food'. She adds that 'identification of thresholds and reversibility of damage' can assist in determining such issues 'in the face of uncertainty' (Carmody 2012: 74).
14. This approach is also broadly consistent with that of Bosselmann, who defines sustainable development as 'the obligation to promote long-term economic prosperity and social justice within the limits of ecological sustainability' (Bosselmann 2008: 53). In this definition, preserving the 'integrity and continued existence of ecological systems' is central (Bosselmann 2008: 53). A failure by current generations to take action to address anthropocentric climate change would clearly violate such an obligation as it threatens the integrity of the global ecological system.
15. But as we shall see in Chapter 5, a human rights approach cannot in itself resolve the distributional dilemmas in climate change policy.
16. Described as 'equal access to atmospheric space' in recent proposals.
17. See also Allen et al. (2009a).
18. Quoted in Singer (2010: 187). (Locke J, (1690) Second Treatise of Government in Macpherson CB (ed.) *Second Treatise of Government* (Hacket Publishing Company Inc 1980)).
19. Referred to in Caney (2011: 97). Caney relies on Zweiniger-Bargielowska (2000: 76, 262, 17), who notes that flat-rate equal allocations of goods were often regarded as unfair but reflected purely practical concerns. Caney uses these examples to argue against equal per capita emissions, whereas my argument is that the rationing example demonstrates how a starting point of equality can be modified, taking into account particular concerns of the most vulnerable and practical concerns in

administration. (Zweiniger-Bargielowska I, *Austerity in Britain: Rationing, Controls, and Consumption 1939–1955* (Oxford University Press 2000)).

20. Caney has pointed out that in relation to climate change, the polluter pays principle is often modified and really embodies a principle of not *benefiting from the harm* caused by pollution given that most historic emissions are produced by people long since dead. This captures the idea that it is *not* just that industrialised countries are primarily responsible for the climate change problem in the first place; it is also that they have primarily benefited from the economic development that comes with high GHG emissions. Caney rejects the polluter pays principle on the basis of the non-identity problem (Caney 2005 [2010]: 128). On the non-identity problem, see Chapter 2.
21. Conceptually of course it is impossible for future generations to consent to actions by current generations; but even a threshold of notional consent based on a future generation consenting in a hypothetical sense on the basis of a cost–benefit analysis would surely fail given the range of irreversible climate change impacts involved and the risk of catastrophic climate change harms.
22. See also Posner and Sunstein (2007: 26).
23. Caney further argues that by ignoring ignorance, one unfairly prioritises the interests of the right bearers whose interests have been affected over the interests of the 'duty bearers' (Caney 2005 [2010]: 132). However, the fact that in national legal systems only some damages are considered 'punitive' implies that in general the intention of compensation is not to punish the agent but to provide a remedy to the victim. Jonathan Pickering helpfully pointed this out.
24. http://australia.gov.au/about-australia/our-country/our-people/apology-to-au (accessed 31 August 2012).
25. See also Shue (1999 [2010]: 105).
26. http://www.unep.org/Documents.Multilingual/Default.asp?documentid=97&articleid=1503 (accessed 31 August 2012).
27. Quoted in Caney (2005 [2010]: 133). (Agarwal A, Narain S and Sharma A, 'The Global Commons and Environmental Justice-Climate Change' in Byrne J, Glover L and Martinez C (eds) *Environmental Justice: International Discourses in Political Economy Energy and Environmental Policy* (Transaction Publishers 2002)).
28. Of course a number of places in schools *at any point in time* is limited, and therefore access to education is in this sense a 'limited resource'. This does not detract from the argument in the text.
29. Thank you to Jonathan Pickering for this insight.
30. An early version of this approach is found in Agarwal and Narain (1991). See also Singer (2002 [2010]: 143), who argues for an approach based on equal per capita future entitlements to a share of the capacity of the atmospheric sink, combined with emissions trading.
31. Contraction and convergence was developed by the Global Commons Institute, see www.gci.org.uk/. See also Page (2006: 177–9).
32. Baer has argued that this approach is unfair to China as it would not allow them the same chance to industrialise as was enjoyed by industrialised countries (Baer 2006 [2010]: 221). However, the idea of giving extra emissions to China (as proposed by Garnaut 2008c) to overshoot per capita emissions may conflict with an effectiveness requirement (see Macintosh 2010).
33. See Miller's (2008: 149) argument to the contrary.

PART II

International law and politics

4. Current international law, intergenerational justice and climate change

4.1 INTRODUCTION

This chapter assesses *current* international law rules in terms of the extent to which they incorporate and satisfy the 'justice principles', *effectiveness imperative* and implementation principles essential for current generations to safeguard the welfare of future generations in relation to climate change set out in the previous chapter. The starting point for this assessment is the current climate change treaty regime, focusing particularly on the UNFCCC and Kyoto Protocol including relevant Conference of Party (COP) decisions up to and including the UNFCCC Doha COP18 held in December 2012 (4.2).[1]

We saw in Chapter 1 that a global treaty is essential for achieving climate change mitigation to ensure justice for future generations owing to trade competitiveness concerns and the need to mobilise technology and finance. An effective global climate regime that meets the imperatives of intergenerational justice requires stringent emission targets and timetables, a funding mechanism to facilitate technology transfer, and a strong compliance and enforcement mechanism (4.2). However, these elements are only found in fragmentary form in the UNFCCC and Kyoto Protocol. The latter only covers the EU and a few other industrialised countries. The Durban mandate for a new global agreement has ambiguities in terms of the legal form of this instrument. Furthermore, while scientists have concluded that global GHG emissions must peak by 2015 and begin declining after that, the Durban mandate only requires a global climate treaty to be negotiated by 2015 with entry into force by 2020 (4.2.1).

The failure to include adequate stringent emission reduction targets in the current global climate change regime has serious flow-on impacts: the existing Kyoto compliance mechanism is limited in scope and effectiveness (4.2.4) and the shape of a compliance mechanism under the global treaty to be negotiated under the Durban mandate remains uncertain.

While the *responsibility for harm* principle entails establishment of a compensation mechanism within the climate change regime, this has been strenuously resisted by industrialised countries. Nevertheless, some movement on this issue occurred at the Doha COP in December 2012, where it was agreed that 'loss and damage' from climate change in poor countries should be addressed, but it was left unstated how much money was involved and, importantly, no acknowledgement of legal liability by industrialised countries for climate-change-related damage was included (4.2.3). The UNFCCC Green Climate Fund represents progress towards an effective financial mechanism which includes within its mandate mitigation, adaptation and technology development and transfer. But financial contributions remain voluntary.

The UNFCCC contains 'intergenerational equity' and 'sustainable development' as guiding principles in Article 3. The reference to intergenerational equity is, however, vague, with no guidance as to how the balance between the needs of present and future generations is to be struck. It thus falls well short of the *core human rights* principle discussed in Chapter 3, according to which persons have equal weight regardless of when they happen to be born.

It is essential to examine general international law to see whether it provides a basis for fleshing out the meaning of 'intergenerational equity', both as a stand-alone principle and as part of 'sustainable development' (4.3.2). Both these principles are embodied in the UNFCCC – along with the 'precautionary principle' and duties of 'cooperation' – which has near universal membership,[2] and as such are binding on practically all states. In addition these principles are important as they are considered by many countries to be integral to the mandate for the current UN climate negotiations under the Durban platform.

It is clear from the following survey that intergenerational equity has no consistent formulation in international environment treaties. Recent ICJ cases show that intergenerational equity and sustainable development have normative force in conjunction with other rules of international law. But the ICJ has tended to focus on the procedural elements of sustainable development in recent cases, such as the *Pulp Mills* case, shying away from pronouncing on the distributive justice elements involved (Stephens 2012: 217). Nevertheless, Justices Weeramantry and Trindade have made strong statements in ICJ cases to the effect that intergenerational equity is now part of customary international law, and these statements may well be built upon in future cases (4.3.1).

International litigation as a means for delivering justice for future generations in relation to climate change faces particular difficulties: how can standing be established for future generations not yet born, and how can damage be demonstrated in advance of its occurring (4.3.2)?

Finally, this chapter examines briefly (4.4) the current procedural rules for the making of international climate treaty law in terms of the requirements of procedural justice for future generations (see Chapter 1). While future generations cannot be represented directly in the UN climate regime law-making processes, representation of their interests through youth NGOs – which largely share their interests – is a possibility, although the potential for effective participation in the current UNFCCC treaty-negotiation process is narrowly circumscribed under current rules.

4.1.1 Sources of International Law

Implicit in the analysis will be a view on the sources of international law and the relationship between international law and ethics. While this book uses a set of justice-based principles as a basis for critiquing international law, this does not imply a sharp distinction between international law and ethics. International law itself has its own ethic/s in the sense of an assumption that 'justice is best served in relations between states rather than individuals' (Glenn 2012: 246). However, this ethic is being transformed with recognition of non-state entities in the realm of human rights and international criminal law (Glenn 2012: 246). A second way in which ethics cannot be avoided in international law is in relation to the sources of international law. While many rules in international law – especially those embodied in treaties – are reasonably clear, rules based on customary international law and 'general principles of law' are not always clear.[3] Whatever doctrine on sources is followed involves – implicitly or explicitly – value judgements which are in essence ethical. Being overly flexible in assessing what qualifies as a rule of international law, in relation for example to 'soft law' instruments which states did not intend to be legally binding, is unhelpful as it can entail that a proposed rule is accepted by states when it is not. This can be a hindrance rather than a help in reforming international law as it allows states to claim that the law has already been reformed to address a particular problem when it has not.

The approach taken here is to take a view on sources of international law resting on the consent of states. This is well expressed by the Permanent Court of Justice in the *Lotus* case:

> International law governs relations between independent States. The rules of law binding upon States therefore emanate from their own free will as expressed in conventions or by usages generally accepted as expressing principles of law and established in order to regulate the relations between these co-existing independent communities or with a view to the achievement of common aims.[4]

While most rules of international law are clear, there are areas of uncertainty. In these areas of uncertainty, it is helpful to be explicit about the value assumptions being made, and to strive for value assumptions likely to assist in international law becoming more certain and fairer.

Sources of international environmental law are particularly difficult for a number of reasons. Firstly, such issues are rarely litigated. Secondly, there is extensive use of so-called 'soft law' referring to political declarations, for example, not intended to be binding under international law. Thirdly, different government ministries take different views on whether a norm is intended to be binding or not depending on the context (Dupuy 2007: 453–5).[5] In spite of these difficulties, the approach taken here is that the rules for identifying whether a particular norm is part of the corpus of international law which apply to international environmental law are no different from general international law (Dupuy 2007: 455).

In referring to environmental norms being created in the framework of international organisations, Dupuy observes that the 'authority of the norm produced' by this process depends, 'regardless of whether it is created through a harder or softer process – upon the conviction of states that the norm has become binding' (Dupuy 2007: 455).[6] The primacy given to states in the sources of international law is not to deny that a range of non-state actors play a role in the law-making process. A wide range of non-government organisations and industry groups have played important roles in the shared understandings underpinning the UNFCCC regime (Brunnee and Toope 2010: 36, 143). State practice also remains an essential ingredient in demonstrating customary international law along with *opinio juris* – the conviction that a rule is legally binding (*North Sea Continental Shelf* cases).[7] However, given the difficulties of demonstrating *opinio juris*, increasing reliance has been placed on conduct within international organisations, including UN resolutions. In a number of cases the ICJ has taken the view that *opinio juris* could be adduced from the attitude of states towards certain UN General Assembly resolutions (for example the *Nicaragua* case[8]), while it was necessary to take into account the content of the resolution and the circumstances surrounding its adoption (Shaw 2008: 88). It is important to observe that, even where there is doubt that a particular principle is binding under international law, it may still have a significant impact on the real world. It may have an impact on expectations, play a role in setting the terms of the debate (Bodansky 1995: 105) or be implemented in national law or impact interpretation of existing international law rules (Beyerlin 2008: 437).

4.2 THE GLOBAL CLIMATE TREATY REGIME: THE UNFCCC AND KYOTO PROTOCOL

You will recall that meeting an *effectiveness imperative* (EI) was proposed as a vital precondition for delivering justice for future generations in the climate change context (Chapter 3). Without an effective regime, future generations will suffer harm contrary to the harm avoidance principle proposed in Chapter 2. Effectiveness is also vital for ensuring that the human rights of future generations are not violated. Thus an effective global climate change regime must meet the objective of: 'stabilization of GHG concentrations in the atmosphere at a level that would prevent dangerous anthropocentric interference in the climate system', derived from Article 2 of the UNFCCC. At the Copenhagen COP in 2009 it was agreed that this objective translates into keeping the earth's temperature below a rise of 2°C (United Nations 2009, Copenhagen Accord para 2). In order to avoid 'dangerous anthropogenic interference with the climate system', climate change scientists are calling for a virtual decarbonisation of the global economy by 2050, with GHG emissions reduced 10–40 per cent by 2020 (see Chapter 1).

We saw above that a global treaty is essential for achieving climate change mitigation owing to competitiveness concerns and the need to mobilise technology and finance (Chapter 1). Such a treaty needs to include the following elements: binding economy-wide mitigation targets, technology transfer-related provisions, including for example a funding mechanism, and an effective compliance mechanism. How the UNFCCC and the Kyoto Protocol embody each of these elements is considered in turn in relation to the effectiveness requirement set out above.

4.2.1 Emission Targets and Timetables

The UNFCCC is a framework convention in the sense that it establishes a number of institutions, including subsidiary bodies. The UNFCCC does not include mandatory mitigation targets and timetables. The relevant provisions have been aptly described as 'soft targets and timetables with many loopholes' (Sands and Peel 2012: 280). Thus Article 4(2) places an obligation on developed country parties and the EC to 'adopt national policies and take corresponding measures on the mitigation of climate change, by limiting its anthropogenic emissions of GHGs and protecting and enhancing its GHG sinks and reservoirs'. But there is no clear and specific commitment to stabilise CO_2 and other GHG emissions at a specific level.

The Kyoto Protocol of 1997 moved beyond the UNFCCC in including specific targets and timetables in relation to six GHGs: CO_2, methane, nitrous oxide, hydrofluorocarbons, perfluorocarbons and sulphur hexafluoride (Annex A). Annex 1 countries – comprising industrialised countries and economies in transition (a number of Eastern European countries) – are subject to specific emission reduction targets (so-called 'emission limitation and reduction commitments') with a view to reducing their overall emissions of the gases covered by the Protocol 'by at least 5% below 1990 levels in the commitment period 2008 to 2012'.[9] No obligations were placed on developing countries, reflecting the principle that industrialised countries should take the lead in reducing emissions. This is a strong reflection of the *responsibility for harm* principle (PII; 3.5.6 above) and *capacity to pay* principle (PIII; 3.5.8 above).[10] Article 2 of the Protocol contains a menu of possible 'policies and measures' which parties may take to achieve their emission reduction goals.[11]

At the December 2009 Copenhagen 15th Conference of Parties (COP) to the UNFCCC, governments agreed on an Accord which contained the 2°C threshold and voluntary commitments by countries to reduce emissions and a mechanism for countries to make further commitments (Copenhagen Accord). The Accord made some headway in breaking down the split between industrialised and developing countries by contemplating both 'quantified economy-wide emissions targets for 2020' by Annex 1 parties and 'mitigation actions' by non-Annex 1 parties (Copenhagen Accord, paras 4–5).[12] Annex 1 party commitments were to be 'measured, reported and verified … in accordance with … Guidelines' adopted by the UNFCCC.[13] Non-Annex 1 parties' mitigation actions, if supported through international financing, would be 'subject to international measurement, reporting and verification', also in accordance with guidelines adopted by the COP.[14] However, the legal status of the

Copenhagen Accord remains contested (Sands and Peel 2012: 294) and in any event it only constitutes a framework for voluntary commitments. Moreover, the voluntary commitments fall well short of delivering what would be required to keep global warming below the 2°C threshold (UNEP 2011).

At COP16 of the UNFCCC held in Cancun, Mexico, the 2°C threshold was captured in a formal Decision. A decision was also made to establish a number of workshops on Equitable Access to Sustainable Development (EASD).

The key decision of the 17th COP to the Kyoto Protocol held in Durban, South Africa in 2011 recognised that climate change represented 'an urgent and potentially irreversible threat to human societies and the planet'[15] and noted 'with grave concern the significant gap between the aggregate effect of Parties' mitigation pledges in terms of global annual emissions of GHGs by 2020 and aggregate emission pathways consistent with having a likely chance of holding the increase in global average temperature below 2°C or 1.5°C above pre-industrial levels'.[16] However, the contrast between this apparent sense of urgency and progress towards an effective global climate regime remains stark.

The Durban COP took a decision to extend the Kyoto Protocol for another five to eight years (2013–2017/20). This decision was followed up by a formal decision to amend the Kyoto Protocol at COP18 held in Doha, 2012.[17] Under this Decision a second commitment period for 2013–20 was agreed. However, the emission targets agreed to will only bind the EU, Switzerland, Australia and a handful of other countries. The US remains a non-party to the Protocol. Canada withdrew in December 2012. Russia, Japan and New Zealand remain parties but did not take on emission targets in this second commitment period.[18] The reduction targets include an EU commitment to reduce its emissions by 20 per cent over this period (2013–20) against a baseline of 1990.[19]

Under this amendment of the Kyoto Protocol, the style of emission targets follows closely that of the first commitment period of the Kyoto Protocol. It therefore retains the same flaw from a justice point of view in that current emissions are used as a reference point (see Chapter 3).

Another key plank in the Durban mandate was the agreement to launch a negotiation process to develop a 'protocol, another legal instrument, or agreed outcome with legal force applicable to all parties' addressing the post-2020 period (hereafter 'Durban Platform').[20] The reference to an 'agreed outcome with legal force' remains unclear, with a concern that this could mean – at least for some of the negotiating parties – an instrument *not* binding under international law (Bodansky 2012a: 1). The Durban conference failed to deliver an agreement on when global emissions should peak or any long-range global emissions reduction target. Moreover, with no language included in relation to the stringency of emissions reductions to be included in the new agreement, the mandate could arguably even be met by another framework treaty (Bodansky 2012a). Of equal concern is the negotiating timetable, with a global agreement to be negotiated by 2015 and 'come into effect and be implemented from 2020'.[21] The Durban mandate included the launch of a 'work plan on enhancing mitigation ambition' aimed at exploring options for increasing the mitigation effort by parties[22] with an in-session workshop to be held at the first negotiating session in 2012.[23]

It is apparent from the outcome of the first such workshop held in Bangkok in August/September 2012 that this 'increased ambition on mitigation' process will not be a backdoor process for renegotiating the Kyoto Protocol or a new global treaty on climate change. Discussion at the Bangkok workshop focused on increasing the number of countries making pledges, the ambition of the existing pledges and recognising additional supplementary actions undertaken 'at subnational, national and international levels'.[24]

Thus in terms of mitigation targets and timetables, the current international legal regime is woefully short of what is required under an *effectiveness imperative* (EI). Stringent emission targets, binding under international law and covering all major emitters, remain lacking. The Copenhagen Accord is only a framework for volunteering action and pledges to date fall well short of what is required to keep warming below 2°C. The mitigation targets of the EU and some other developed countries under the Kyoto Protocol reflect the *responsibility for harm* and *capacity to pay* principles. However, without binding emission targets for all industrialised countries and major developing country emitters, the *effectiveness imperative* is not met, and consequently the core human rights of future generations are threatened (PI; Chapter 3). Moreover, the current negotiating mandate under the Durban Platform is inadequate, owing to soft time frames and a lack of stringency built into the mandate itself.

4.2.2 Funding Mechanisms and Technology Transfer

Adequate funding and technology transfer arrangements are also essential components required for a global climate change regime to meet the *effectiveness imperative* (see Chapter 1). The UNFCCC included an obligation (in Article 4(3)) on developed country parties to 'provide such financial resources, including for the transfer of technology, needed by the developing country Parties to meet the agreed full incremental costs of implementing' the general obligation set out in Article 4(1) (referred to above) and that are agreed between a developing country party and the financial mechanism to be decided upon under Article 11 of the Convention.[25] While of enormous potential impact, this principle has not yet been implemented owing to the lack of appropriate mitigation targets and timetables.

The Copenhagen Accord included a commitment by industrialised countries to provide 'new and additional resources approaching USD30 billion for the period 2010–2012' balanced between adaptation and mitigation with the longer-term collective goal of mobilising USD100 billion per year by 2020 'to address the needs of developing countries' from a combination of sources including public, private, bilateral and multilateral, but this money will be linked to 'meaningful mitigation actions and transparency and implementation'.[26]

The Durban COP made good progress in terms of thrashing out most of the governance arrangements relating to this fund, allowing it to be launched.[27] The fund is to balance adaptation and mitigation.[28] The Green Climate Fund is to be based on voluntary financial inputs from developed country parties but also includes public and private sources.[29] Developing country parties to the Convention are eligible to access the fund, which will finance 'agreed incremental costs' for activities relating to 'adaptation, mitigation, … technology development and transfer, capacity building and the preparation of national reports by developing countries' as well as REDD-plus (relating to reducing emissions from deforestation in developing countries).[30]

In summary, then, the Green Climate Fund represents progress towards an effective funding mechanism but its potential effectiveness is weakened by financial contributions remaining voluntary. The lack of appropriate emission reduction targets also weakens the possibilities for funding developing country mitigation efforts and facilitating the technology development and transfer required.

4.2.3 Compensation

The *responsibility for harm* principle (PII; Chapter 3) entailed the idea that a person who caused harm should clean up the resulting harm and/or provide suitable compensation. This principle has been partly reflected in the structure of mitigation targets in the climate regime.

However, within the UNFCCC (Kyoto Protocol) regime, the idea of establishing a compensation mechanism for climate-change-related damage until very recently has not gained any political traction. Industrialised countries have been concerned at the costs involved. Moreover, they consider it unfair for them to be held responsible for emissions during the period when they did not realise that they were causing damage. Thus a proposal to include the polluter pays principle in the UNFCCC was rejected (Slade 2007: 218).[31] Similarly, a proposal to confer on the UNFCCC Compliance Committee power over a state to pay for restoration of damage to the environment was not accepted (Lefeber 2012: 328). Reflecting this attitude, the US delegate stated at the 2009 Copenhagen Summit on Climate Change that '[w]e absolutely recognize our historic role in putting emissions in the atmosphere, up there, but the sense of guilt or culpability or reparations, I just categorically reject that.'[32]

The debate on this issue, however, does seem to be gradually shifting in a direction reflecting, at least to some extent, the responsibility for harm principle. Thus, at the Doha COP in December 2012, it was agreed that 'loss and damage' from climate change in poor countries should be addressed.[33] While the decision requests countries to provide to this end 'finance, technology and capacity building' (para 8), it is left unclear how much funding is involved and where it should come from. Crucially, the decision contains no acknowledgement of legal liability. US agreement was contingent on funding being drawn from existing commitments (Jotzo and Pickering 2012: 1).

4.2.4 Compliance and Enforcement

In order to be effective, mitigation targets need to be linked to an effective compliance mechanism. The Kyoto Protocol contains a complex enforcement and compliance mechanism (Brunnee 2012: 290–320).[34] Non-compliance by parties involves the possibility – ultimately – of a suspension of a party from mechanisms of the Protocol, including emissions trading. The Protocol's enforcement and non-compliance mechanism has had mixed success to date, playing a positive role in facilitating compliance by some parties but proving powerless in relation to Canada's open refusal to meet its obligations under the Protocol (Brunnee 2012: 309–10).

Ultimately, collective compliance and enforcement are dependent upon agreement on international standards (Brunnee 2012: 320). The Kyoto compliance and enforcement mechanism will continue in relation to the recently agreed second Kyoto Protocol commitment period, although as we have seen, this covers a more limited range of countries. Whether an effective compliance procedure is included in the global instrument which results from the Durban Platform remains uncertain. Already in the recent Bangkok meeting, widely diverging views were expressed on the form which such an enforcement/compliance procedure should take.[35]

Ultimately, an effective enforcement and compliance mechanism and a funding mechanism remain vital ingredients in an effective global climate change regime. Both of these elements remain highly dependent upon agreement on clear and binding emission reduction commitments in a global climate treaty.

4.2.5 Duty of Cooperation

Chapter 3 argued that there is an ethical obligation on states to cooperate in creating an effective international climate change regime. At a general level, international environmental law already includes such an obligation (Sands and Peel 2012: 203). This obligation rests on the UN Charter Article 74 and has been affirmed in a number of ICJ cases, including the *Pulp Mills* case.[36] The UNFCCC and Kyoto Protocol include specific obligations to cooperate. The UNFCCC in its Preamble acknowledges that:

> the global nature of climate change calls for the widest possible cooperation by all countries and their participation in an effective and appropriate international response, in accordance with their common but differentiated responsibilities and respective capabilities and their social and economic circumstances.[37]

The UNFCCC places an obligation on parties to cooperate: in the development and transfer of technology; in adaptation, in scientific research on climate change, in the exchange of information (socio-economic, legal and technical); and in education, training and public awareness on climate change.[38] A difficulty with these obligations to cooperate embedded in the UNFCCC is that it is almost impossible to judge whether a particular party has met these obligations owing to their open-ended nature.

4.2.6 Precautionary Principle

Environmental effectiveness requires acceptance of a precautionary approach (Chapter 3). An ethical obligation to take a precautionary approach is reflected in Article 3 of the UNFCCC, which provides that parties:

> should take precautionary measures to anticipate, prevent or minimize the causes of climate change and mitigate its adverse effects where there are threats of serious or irreversible damage, lack of full scientific certainty should not be used as a reason for postponing such measures, taking into

> account that policies and measures to deal with climate change should be cost-effective so as to ensure global benefits at the lowest possible cost.

As discussed above, there is more than sufficient scientific certainty to justify strong mitigation action on climate change and therefore the precautionary principle is strictly not applicable. The precautionary principle is closely linked to a *preventive* approach whereby action should be taken prior to environmental impacts becoming irreversible. Seen in this light, the precautionary principle is valuable in underscoring the need for urgent mitigation action on climate change. However, it is vague in not specifying what measures are required. While the precautionary principle is well established as a principle of EC law (Scotford 2008: 37–40), there remains some uncertainty as to whether it is part of customary international law binding on all states (ILA 2012: 21).[39] The uncertainty as to the status of the precautionary principle in general international law is irrelevant for our purposes, because parties to the UNFCCC have accepted that the principle applies in relation to climate change.

4.2.7 Intergenerational Equity

The UNFCCC also contains a reference to intergenerational equity. Article 3 provides that 'in their actions to achieve the objective of the Convention and to implement its provisions, the Parties shall be guided, inter alia, by the following:

> the Parties should protect the climate system for the benefit of present and future generations of human kind, on the basis of equity and in accordance with their common but differentiated responsibilities and respective capabilities. Accordingly, the developed country Parties should take the lead in combating climate change and the adverse effect thereof.

What is the status of the principles contained in Article 3? During the negotiation of the UNFCCC, the US and some other countries vehemently opposed inclusion of the principles contained in Article 3, concerned that these principles would be invoked against them under the Convention's dispute settlement provisions (Bodansky 1993: 501). These principles embody 'legal standards' of a more general nature, which have normative force, rather than commitments which 'specify particular actions' (Bodansky 1993: 501).[40] The principles are inherently vague. For instance, while 'equity' is to apply in relation to the protection of the climate system 'for the benefit of present and future generations', no guidance is given as to how the balance between the needs of present and future generations is to be struck. Thus it remains unclear whether the reference to 'intergenerational equity' reflects the notion of persons having equal weight regardless of when they happen to be born, integral to the *core rights principle* set out in Chapter 3 above.

4.2.8 Common but Differentiated Responsibilities

The reference to developed country parties taking 'the lead in combating climate change' is a reflection in a general sense of the justice principles argued for in Chapter 3. However, it is ambiguous in that it remains unclear which of these justice principles it reflects. This ambiguity reflects the different approaches taken by parties to the negotiations. Developing countries argued that the notion of industrialised countries taking the lead in addressing climate change reflected the responsibility of these countries for creating the climate change problem. In contrast, industrialised countries took the view that it reflected their greater 'financial and technical capacity' and denied the suggestion of responsibility for historic emissions (Bodansky 1993: 503). The reference to 'respective capabilities' in Article 3 reflects the 'capacity to pay' principle argued for (PIII; Chapter 3).

Both the Preamble and Article 3 of the UNFCCC include the concept of 'common but differentiated responsibilities' (CBDR). This concept has two elements: the notion of common responsibility in implementing an agreement and the notion of states having differentiated obligations based on their respective roles in creating a particular environmental problem and capacity to solve the problem (Rajamani 2006: 134, 136).

The wording of Article 3 leaves unclear whether the basis of differentiation between countries is with respect to their capabilities or responsibility for causing the problem (Rajamani 2006: 194–5).[41] The reference to CBDR and industrialised countries' leadership in addressing climate change are couched in 'discretionary' and 'guiding language' rather than hard legal obligations, leading Rajamani to conclude that they are not legally binding or part of customary international law.[42] However, both concepts have considerable 'force' in the climate regime, both as a basis for the interpretation for existing obligations and providing an 'overarching principle governing the future development of the climate regime' (Rajamani 2006: 197).[43] Similarly it can be argued that the 'polluter pays principle' (3.5.6) as an economic policy or ethical principle finds expression in CBDR as a principle of interpretation (Buchner and Lehmann 2005: 55).

As we saw above, the 'differentiation' element of CBDR is embodied in both the UNFCCC and Kyoto Protocol in that the 'quasi target' found in Article 4(2)(a) and (b) of the UNFCCC only imposes obligations on industrialised countries and the Kyoto Protocol emissions reduction targets only apply to industrialised countries.

However, CBDR in its current form in the global climate regime arguably does not reflect the justice principles of *responsibility for harm* and *capacity to pay* as the current climate regime does not impose mitigation obligations on large developing country emitters such as India, China and Brazil (Atapattu 2009: 42).[44]

Surprisingly, the mandate for a new global legal instrument under the Durban platform has no reference either to equity or to CBDR. This is consistent with the US position, which equates CBDR with the notion of 'major emitter' developing countries not taking on binding emission reduction commitments. The US has consistently maintained that 'major emitter' developing countries along with developed countries must be included in any global climate treaty (Pickering et al. 2012: 431).

Nevertheless, many developing countries have made clear that they understand both equity and CBDR as integral to the Durban mandate. Thus at the Doha COP in 2012 a statement made by China, Brazil, India and South Africa (the so-called BASIC group) reiterated that 'equity' and CBDR should 'guide future negotiations'.[45]

4.2.9 Subsistence and Equality

The UNFCCC does not directly incorporate the principle of subsistence discussed in Chapter 3. However, the Preamble refers to 'the legitimate priority needs of developing countries for the achievement of sustained economic growth and the eradication of poverty'.[46] Moreover, at the UNFCCC COP16 held in 2011 in Cancun, Mexico, it was agreed that Parties should cooperate in achieving a peaking of global GHGs 'as soon as possible … bearing in mind that social and economic development and poverty eradication are the first and overriding priorities of developing countries'.[47]

Some justice principles and implementation principles argued for in Chapter 3 are notably absent from the current global climate regime. For example, there is no reference to the notion of 'equality of access to the atmosphere', or 'equal per capita emissions' beyond the reference in the UNFCCC Preamble to 'per capita emissions in developing countries' being still 'relatively low' while the 'largest share of historical and current global emissions of GHGs has originated in developed countries'.[48] However, the UNFCCC does include a reference to 'sustainable development' (Article 3(4)) and it is to this concept that we now turn.

4.2.10 Sustainable Development

The UNFCCC states that parties have a 'right to, and should, promote sustainable development'.[49] This was a compromise between developing countries wishing inclusion of a *right* to development and some industrialised countries wishing to see inclusion of a *duty* on states to promote sustainable development (Bodansky 1993: 504). The US in particular was strongly opposed to a 'right to development' as it was considered too vague and could be used by developing countries to demand financial assistance from developed countries.

The reference to a 'right' to 'promote sustainable development' rather than the more usual 'duty' of sustainable development is arguably different from the right to development found in the 1986 UN Declaration on the Right to Development (Bodansky 1993: 504). It is also different from the Rio Declaration, which states that 'the right to development must be fulfilled so as to equitably meet developmental and environmental needs of present and future generations'.[50] Given the failure of the UNFCCC to define 'sustainable development' (ILA 2010: 25), it is important to see whether general international law sheds light on the meaning of this principle as embodied in the UNFCCC.

4.3 GENERAL INTERNATIONAL LAW

4.3.1 Principles of International Environmental Law: Intergenerational Equity and Sustainable Development

Intergenerational equity arises both as an independent principle and also as a component part of sustainable development. It is important to consider the provisions relating to intergenerational equity in the UNFCCC in the context of general international law for two reasons. Firstly, this may be relevant to the interpretation of intergenerational equity as it is embodied in Article 3 of the UNFCCC. Secondly, it is important to ascertain whether there is a general obligation of intergenerational equity in international law which is binding independently of the UNFCCC treaty.

Rules of international law can emerge through repetition of a particular rule in a number of multilateral treaties, provided certain conditions are met. These conditions include that there be consistency in expression of the rule and that the parties to the treaty recognise the rule to be of binding force on all states (*North Sea Continental Shelf* cases).[51]

Intergenerational equity has been embodied in global environment treaties in various forms. The 1992 Convention on Biological Diversity[52] contains a vague reference to the sustainable use of biological diversity to meet the needs and aspirations of 'present and future generations' but contains no obligations to guarantee continuation of habitat essential to the survival of endangered species. The 1992 Rio Declaration on Environment and Development states: 'the right to development must be fulfilled so as to equitably meet development and environmental needs of present and future generations'.[53] As in Article 3 of the UNFCCC, intergenerational equity is intertwined with intragenerational rights.

Another expression of intergenerational equity discourse involves conserving particular elements of nature for the benefit of future generations. This approach is embodied in the 1972 World Heritage Convention,[54] which aims to conserve the cultural and the natural heritage of 'outstanding universal value to the benefit of future generations'.[55] Finally, it should be noted that there are a series of important international environmental treaties that have not explicitly referred to intergenerational equity but have nevertheless been mindful of the concept. Thus several of the treaties within the Antarctic Treaty System refer to management of Antarctica 'in the interest of all mankind', implying a responsibility to future generations (Bastmeijer 2011: 2–3).[56] A further example is the 1987 Montreal Protocol on Substances that Deplete the Ozone Layer,[57] which aims to reduce the impact of ozone-depleting substances now to ensure a restored ozone layer in the future.[58] Like most treaties, however, the overriding imperative of the Montreal Protocol is to protect the interests of current generations' health and access to food resources,[59] and it is based on a belief in human capacity to redress harm before it is too late.

There is no single formulation of intergenerational equity in these treaties spanning different areas of the environment, so it is difficult to argue that the principle has a sufficiently consistent meaning in general international law capable of informing the interpretation of Article 3 of the UNFCCC. Indeed, on one view the references to 'intergenerational equity' in global environment treaties has tended to be 'hortatory', not even having weight as 'principles' let alone 'rules' of international law (Fitzmaurice 2009: 128–9).

Article 3 of the UNFCCC links intergenerational equity to a right to sustainable development. Can intergenerational equity be considered part of customary international law as a component of sustainable development? To establish a rule of customary international law it is necessary to point to consistent state practice and *opinio juris*. A further requirement is that a rule be of a 'fundamentally norm creating character'. This latter requirement means that a rule must be sufficiently clear (*North Sea Continental Shelf* cases).[60]

There is extensive state practice in the form of a growing number of states incorporating sustainable development, including intergenerational equity, in their national laws and constitutions.[61] Examples include South Africa's National Environmental Management Act[62] and the Australian case *Taralga Windfarm*, which required a planning decision in relation to a windfarm to take into account future generations' needs for renewable energy in the context of climate change.[63] While there may be sufficient state practice, there is insufficient *opinio juris* demonstrating that states have implemented sustainable development accompanied by a belief that there was an international legal obligation to do so, as opposed to a political or policy commitment (Cordonier Segger 2008: 141).

Moreover, both intergenerational equity and sustainable development fail to meet the test of being of a 'norm creative character' because they are too vague and indeterminate. Intergenerational equity does not in itself indicate what weight is to be given to the interests of future generations vis-à-vis current generations. Similarly, sustainable development does not indicate the respective weight to be given to environmental, particularly long-term, interests vis-à-vis short-term economic interests. As pointed out by Birnie, Boyle and Redgwell, even if intergenerational equity is viewed as an element of sustainable development, this 'does not resolve the argument for stronger generation rights or international guardianship, nor does it determine the optimal balance between this generation and its successors' (Birnie et al. 2009: 122).

Even if sustainable development is not a rule of customary international law, it still can have normative force in international law. Indeed, an International Law Association (ILA) Committee has argued that, while the legal status of sustainable development is a 'sterile' issue, the practical effect of sustainability as a *concept* is more important (ILA 2012: 6). Thus sustainable development can act as a catalyst for the further development of international law and a vehicle for bridging the North–South divide (Beyerlin and Marauhn 2011: 79–83). More concretely, sustainable development can have normative force in international law by: (1) being the object and purpose of a treaty (Cordonier Segger 2008: 142); (2) being used in the interpretation of treaties; and (3) operating as principles in mediating between other rules or principles of international law.[64] This normative force is reflected in various ways in cases which have come before the ICJ.

The International Court of Justice (ICJ) made references to intergenerational issues in its *Advisory Opinion on the Legality of the Use of Nuclear Weapons*,[65] where the Court considered a request from the United Nations

General Assembly on the question of whether the threat or use of nuclear weapons in any circumstances was permitted under international law. The majority judgment, in the context of considering the relevance of international environmental agreements relating to times of peace, stated that '[t]he court also recognises that the environment is not an abstraction but represents the living space, the quality of life and the very health of human beings, including generations unborn.'[66] The Court, however, went on to argue that obligations stemming from environmental treaties could not have 'intended to deprive the state of the exercise of its right of self-defense under international law', suggesting that environmental considerations needed to be taken into account by states 'when assessing what is necessary in proportion to the pursuit of legitimate military objectives'.[67] However, the Court did not rely directly on the impact of the use of nuclear weapons on future generations in its judgment. As observed by Sands and Peel (2012: 210), 'the purpose of the ICJ's reliance on the intergenerational equity concept is not immediately apparent'.

In his dissenting opinion, Judge Weeramantry took a very different approach, concluding that the threat or use of nuclear weapons was 'illegal in any circumstances whatsoever'.[68] After noting that the half-life of the radioactive elements of a nuclear explosion could run to tens of thousands of years, he stated that 'at any level of discourse, it would be safe to pronounce that no one generation is entitled, for whatever purpose, to inflict such damage on succeeding generations'.[69] He added that the court in applying international law must 'pay due recognition to the rights of future generations', adding '[i]f there is any tribunal that can recognise and protect their interests under the law, it is this court.' Judge Weeramantry went on to say that:

> the rights of future generations have passed the stage when they were merely an embryonic right struggling for recognition. They have woven themselves into international law through major treaties, through juristic opinion and through general principles of law recognized by civilized nations.[70]

Justice Weeramantry's judgment certainly provides valuable material for arguing intergenerational equity in future cases before international tribunals. However, it is regrettable that his judgment does not flesh out more fully the basis upon which future generations can be bearers of rights, how exactly intergenerational equity can be based on 'general principles of law', and how multilateral treaties which enshrine intergenerational equity can create obligations on all states.

In the case concerning the *Gabcikovo-Nagymaros Dam* (Hungary/Slovakia) 1997, the ICJ considered a dispute between Hungary and

Slovakia relating to the construction of a hydro-electric facility by Hungary which impacted the environment of downstream Slovakia.[71] The majority judgment made reference to 'new scientific insights and to a growing awareness of the risks for humankind – for present and future generations – of pursuit of interventions in nature'. The judgment went on to recognise the development of 'new norms' which needed 'to be taken into consideration … not only when states contemplate new activities but also when continuing with activities begun in the past. This need to reconcile economic development with protection of the environment is aptly expressed in the concept of sustainable development.'[72]

This passage is ambiguous, leaving it unclear whether these new norms 'bind as rules of law' and whether sustainable development has indeed become part of general international law (Lowe 1999: 20).

Nevertheless, sustainable development is integral to the Court's order for the parties to go away and negotiate a solution to the volume of water issue, presumably balancing environment and economic considerations (Sands and Peel 2012: 208). Indeed, sustainable development – when functioning as a consideration influencing a particular decision in a particular direction – fits well within Dworkin's notion of 'legal principles'. According to Dworkin, legal principles do not dictate a particular outcome in the way in which a legal rule does but still retain normative force as principles are taken into account in reaching outcomes.[73]

The majority's judgment contrasts with Judge Weeramantry, who in dissent considers that sustainable development is a legal principle of customary international law with an *erga omnes* character based on its general acceptance by the international community, repeated in many instruments and also by 'logical necessity' required to balance the right to development with the need for environmental protection. He finds support for sustainable development in material derived from a range of cultures and legal systems over thousands of years.[74]

Intergenerational equity was further considered by Judge Weeramantry in his dissenting opinion in the 1995 second phase of the *Nuclear Tests* case involving New Zealand and France.[75] In this case New Zealand argued that the basis of the 1974 judgment by the ICJ had been affected in a manner which justified a reopening of the proceedings.[76] New Zealand argued that France, by continuing underground nuclear testing in the Pacific, would violate New Zealand's rights under international law, and in the alternative that such testing would be unlawful without France first conducting an environmental impact assessment. The Court rejected New Zealand's argument on the grounds that the 1974 judgment was based on atmospheric testing and so was not 'affected'. Judge Weeramantry in his vigorous dissent referred to what he described as 'the concept of intergenerational rights', which he described as 'an important and rapidly developing principle of contemporary environmental law'.[77] After referring to the information presented to the Court indicating that 'the half life of a radioactive by-product of nuclear tests can extend to over 20,000 years', he went on to argue that the ICJ should consider *itself* a trustee of the rights of future generations:

> … this Court must regard itself as a trustee of those rights in the sense that the domestic court is a trustee of the interests of an infant unable to speak for itself. If this court is charged with administering international law, and if this principle is building itself into the corpus of international law, or has already done so, this principle is one which must inevitably be a concern of this court. This consideration involved is too serious to be dismissed as lacking in importance merely because there is no precedent on which it rests.[78]

Justice Weeramantry further stated that New Zealand's complaint related not just to the rights of people presently in existence but also 'the rights of unborn posterity. Those are rights which a nation is entitled, and indeed obliged, to protect.' He concluded that this aspect was 'not to be ignored' in considering whether New Zealand had made out a prima facie case of damage to its interest sufficient to reopen the 1974 proceedings. Justice Weeramantry stated that the present case provided the Court with a 'pre-eminent opportunity' to make a pronouncement on this developing field, as 'it raises in pointed form the possibility of damage to generations yet unborn'.[79]

Other ICJ judges have not taken up Justice Weeramantry's novel idea of the ICJ being directly a trustee for the rights of future generations. Perhaps this is not surprising as surely for the Court to be a trustee in this fashion would require either that this role be spelled out in the Court's statute which establishes its jurisdiction, or be specified in treaties which the Court is applying through Article 38 of its statute.

While we have seen that there are obstacles in establishing an obligation of intergenerational equity under customary international law, there is no doubt that the concept, particularly as an integral part of sustainable development, has begun to play an important role in the reasoning of international tribunals. Lowe argues convincingly that sustainable development is an element of judicial reasoning, constituting:

> a meta principle, acting upon other legal rules and principles – a legal concept exercising a kind of interstitial normativity, pushing and pulling the boundaries of the primary norms when they threaten to overlap or conflict with each other (Lowe 1999: 31).

The Arbitration Tribunal operating under the auspices of the Permanent Court of Arbitration in the 2005 case of *Iron Rhine (Belgium v Netherlands)*[80] suggested that sustainable development required states to take environmental protection considerations into account in the development process (Cordonier Segger 2008: 126). However, the arbitral tribunal, by emphasising integration of economic and environmental aspects as a touchstone of sustainable development, left unclear the extent to which intergenerational equity was an essential part of this concept.

Intergenerational equity and sustainable development were invoked by both Argentina and Uruguay in their recent conflict, which resulted in the ICJ *Pulp Mills* decision of 2010. The ICJ in this case decided that Uruguay had violated various procedural obligations in authorising construction of two pulp mills on the River Uruguay, a shared water course. These obligations, which included duties to inform the other party of construction plans which interfered with the river, were contained in a bilateral treaty between the two countries, the 1975 Statute of the River Uruguay.

Uruguay emphasised sustainable *economic* development, arguing that it had a right to sustainable economic development that implied a right to construct the pulp mill. In its pleadings it maintained that sustainable development 'is a matter of intergenerational equity, requiring that economic development proceed in a manner that it regulates protection of the environment, which is the human life-support system on which both present and future generations depend'.[81] In contrast, Argentina argued in its oral arguments that 'an effective application of the principles of prevention and precaution by Uruguay was necessary to apprehend the risk of grave damage for present and future generations'.[82]

However, the majority judgment in this case was based almost entirely on interpretation of the 1975 statute rather than principles of international environmental law, such as sustainable development or intergenerational equity. Thus the Court made reference to provisions in the 1975 statute which provided for the balancing of the riparian states' interests in the river as being 'consistent with the objective of sustainable development'.[83] The Court went on to state that Article 27 of the statute which contained obligations relating to the use of the river embodied 'this interconnectedness between equitable and reasonable utilisation of a shared resource and a balance between economic development and environmental protection that is the essence of sustainable development'.[84]

The relatively thin reference to sustainable development or intergenerational equity by the majority judgment was criticised strongly by Justice Trindade in his separate opinion.[85] He chided the majority judgment for not referring to intergenerational equity in interpreting the 1975 statute requirements relating to monitoring.[86]

Trindade addressed intergenerational equity at some length. His starting point was the notion of 'conservation' which reflects 'a cultural manifestation of the integration of the human beings with nature and the world where-in he or she lives'.[87] This is linked to an obligation to other generations 'past and future'. This obligation is in turn linked to a 'preventative (and precautionary) character' in environmental protection[88] to other areas of international law such as human rights which extend obligations into the future to 'potential or prospective victims'.[89]

Trindade goes on to argue that the 'preventive' element of both international human rights and environmental law serves as an 'inspiration for the progressive development of International Law'.[90] He notes that a concern for future generations underlies various treaties and instruments, including the 1977 UNESCO Declaration on the Responsibilities of the Present Generations towards Future Generations[91] and the UN Human Rights Covenants. He quotes extensively from the 1988 Guidelines on Intergenerational Equity[92] – which he took part in drafting – which constitute recommendations on intergenerational equity and international law, strongly influenced by Brown Weiss's notion of intergenerational equity, including the idea of conveying the earth to future generations in no worse condition than it has been received.[93] Trindade notes that intergenerational equity has 'an essentially anthroprocentric outlook'.[94] After referring to the precautionary principle, he concludes that, 'in 2010, it can hardly be doubted that … intergenerational equity forms part of conventional wisdom in International Environmental Law'.[95]

Trindade places importance in 'solidarity in time' as integral to intergenerational equity in the sense of preservation of culture from the past to the future. He refers at length to the Inter-American Court of Human Rights (IACtHR) 2001 decision in *Community Mayagna (Sumo) Awas Tingni*[96] to protect the communal property of an indigenous group in Nicaragua.[97] He observed that 'living in harmony with their natural environment' was an essential part of their culture, and the IACtHR took into account the relationship to the land which had a 'spiritual element' essential for transmitting their cultural legacy to future generations.[98] He goes on to link this notion of cultural identity to the right to life.[99]

Trindade discusses sustainable development, describing its endorsement in various treaties and declarations.[100] He argues that sustainable development is a general principle of international environmental law 'acting as a guiding general principle for the consideration of environmental and developmental issues'.[101] However, he does not explain how sustainable development or intergenerational equity should specifically have been taken into account in interpreting the 1975 Statute given that these principles were invoked by both sides.

The inherent pliability of these principles is commented on by Justices Al-Khasawneh and Simma in their joint dissenting opinion.[102] They point out that the principles of permanent sovereignty over natural resources, and other principles including the principle of sustainable development, exhibit an 'extreme elasticity and generality'.[103] Moreover, they point out that these principles are frequently 'where there is a dispute, in a state of tension with each other'.[104]

It remains to be seen whether Trindade's strong support for intergenerational equity and sustainable development as comprising general principles of international law has influence. His reference to intergenerational equity as 'part of conventional wisdom' is obscure: it is unclear whether he is concluding that intergenerational equity is an underlying objective of international environmental law or part of international law established by repetition in treaties. Interestingly, Trindade's notion of sustainable development is anthropocentric, with a strong emphasis on human beings' harmony with nature. He blurs the distinction between these principles as ethical and policy prescriptions and international law. However, both sustainable development and intergenerational equity in his view have normative force in the sense that they should have been taken into account in interpreting the 1975 statute. His reference to 'progressive development' of international law suggests that he wishes to see international law incorporate these principles, with the ICJ playing an active role in this process.

The ICJ will have such an opportunity to further develop the concept of intergenerational equity in deciding the dispute involving Australia and New Zealand's challenge to Japan's whaling programme. At the time of writing, the court was hearing submissions by the parties. This case involves the issue of whether Japan's current whaling programme is justified under the International Whaling Convention (ICRW) provision for scientific whaling. An Australian submission to the Court argued that the ICRW should be interpreted in light of the development of international environmental law since the ICRW was drafted in 1946. The submission went on to state that there were three juridical pillars recognised as essential for international environmental law, comprising intergenerational equity, the principle of prevention and the precautionary principle. Attention was drawn to the specific reference to intergenerational equity in the Preamble of the ICRW as well as other international instruments and jurisprudence of the ICJ which placed an emphasis on the necessity of taking into consideration the rights of future generations.[105]

While sustainable development and intergenerational equity have substantive notions of fairness and justice embedded in them, international courts have tended to shy away from engaging in broader notions of justice entailing 'equitable distribution among peoples of the public goods that the environment provides' (Stephens 2012: 202).[106] Rather than engage in the substantive distributional element of sustainability, international courts have focused on procedural obligations (Stephens 2012: 217). An example of this is the focus on environmental impact assessment in the *Pulp Mills* case, where the Court concluded that Uruguay's obligations under the 1975 statute entailed the carrying out of an environmental impact assessment (Stephens 2012: 214).[107]

We have seen from recent ICJ judgments that sustainable development and intergenerational equity have normative force in conjunction with other rules of international law. However, the dependence of international dispute resolution bodies on state consent has meant that such bodies have been reluctant to develop international law beyond certain limits (ILA 2012: 7). Reflecting this political reality, the ICJ to date has been 'wary' of pronouncing on the 'formal status of sustainable development' or identifying 'any concrete demands that sustainable development makes of State behaviour' (ILA 2012: 9).

This reluctance means that ICJ jurisprudence does not assist greatly in the interpretation of 'intergenerational equity' in Article 3 of the UNFCCC. Moreover, the inherent indeterminacy of both intergenerational equity and sustainable development means that they are incapable in themselves of establishing clear climate change mitigation burdens or addressing the issue of how to distribute mitigation burdens between current and future generations. For the latter, more precise mitigation rules need to be anchored in international treaty obligations, reflecting the justice principles set out in Chapter 3.

4.3.2 State Responsibility and International Litigation

This section considers whether international law rules on state responsibility can be used as a basis for claims made bilaterally or before international tribunals based on breaches of international law obligations requiring climate change mitigation in ways which satisfy the justice principles set out in Chapter 3. Human rights litigation is considered separately below.

International law rules on state responsibility could arise in the context of bilateral claims, between a small island state threatened with inundation and a major GHG emitter, or in the context of international litigation before an international tribunal such as the ICJ. The object of such claims or litigation could include reparation for previous climate-change-related damage or action to compel state/s to take mitigation action.

While reparations would be consistent with the *responsibility for harm principle*, they could only partially fulfil the requirements of the justice principles owing to their backward-looking and ad hoc nature. The potential link between reparations and future generations is limited, as, in order to benefit future generations, compensation would need to go into, for example, a fund devoted to climate change mitigation action, or be used as an incentive for large emitters to reduce emissions.

To establish state responsibility for an internationally wrongful act, it is required to demonstrate a breach of an international law obligation resulting from conduct attributable to a state.[108] In the context of climate change it would be necessary to show breach of a treaty or customary international law obligation. As we have seen, to date, the only clear and binding emission mitigation targets in a global climate treaty are the obligations on Annex 1 countries under the Kyoto Protocol. The UNFCCC contained vague 'quasi-targets' for developed countries. Voigt (2008: 6) makes an argument that state parties to the UNFCCC have an obligation to take measures to stop increasing their GHG emissions by combining Articles 2 and 4.2 of the Convention. But she acknowledges that this view is 'controversial' and 'no consensus exists' on this interpretation (Voigt 2008: 7). The voluntary commitments made by states in 2010 at the Cancun COP were not binding (Lefeber 2012: 325). Under the UNFCCC, developing countries have been invited to adopt mitigation measures but 'have no binding obligations to this end'.[109]

The Kyoto Protocol targets do impose strict obligations on industrialised countries, binding under international law, to be achieved by the end of 2012. However, non-performance of these obligations, which relate to the first commitment period 2008–12, will only be first assessed in 2015 following the Kyoto Protocol's reporting and review procedures, and even then, countries in breach could escape liability by purchasing carbon credits on international markets (Brunnee 2012: 305). As we have seen, the second commitment period of the Kyoto Protocol (2013–20) only involved the EU, Australia and a limited number of countries.

A further possibility would be to make a claim bilaterally or through international litigation based on a breach of the customary international law obligation not to produce transboundary environmental damage. This rule requires states 'to ensure that activities within their jurisdiction and control protect the environment of other states or of areas beyond national control' (*Nuclear Weapons* case).[110] This principle is binding on all states and thus in theory could be invoked against a state such as the

US which is not a party to the Kyoto Protocol (but is a party to the UNFCCC). However, such a claim would face a number of obstacles. While it is well established that potential damage must be significant to invoke the principle, it is less clear what methodology should be used for working out the threshold of damage (Lefeber 2012: 340). A further obstacle includes establishing causation in a context where all states contribute to GHG emissions and damage has also been caused by historic emissions. However, one solution would be to assign 'proportionate responsibility' established using data compiled under the UNFCCC (Lefeber 2012: 346). Causation would still arguably be difficult as it would be necessary to untangle natural and anthropogenic causes – storm surges occur naturally in Bangladesh but will become more severe and frequent resulting from climate change (see Chapter 1).

While it may be possible to establish jurisdiction before an international tribunal,[111] the prospect for compliance with any ruling would be slim. Given that, to date, states have refused binding and deep mitigation targets in a treaty, it is difficult to see why they would accept such requirements through litigation or pressure pursuant to a bilateral claim.[112] But such claims or litigation could increase political pressure on policy-makers by raising the profile of the climate change issue, as has arguably occurred through litigation at the national level in the US (Osofsky 2012: 455) and in other countries (Lord et al. 2012).

4.4 PROCEDURAL JUSTICE AND THE INTERNATIONAL CLIMATE REGIME

Future generations cannot of course directly participate in the global climate change treaty-making processes. However, future generations' interests significantly coincide with the interests of young people. Young people may be represented as part of national delegations. Youth climate NGOs participate in the UN climate negotiations as observers. The UNFCCC provides that: '[a]ny body or agency, whether national or international, governmental or non-governmental, which is qualified in matters covered by the convention' may apply for an accreditation as an observer.[113] The UNFCCC Secretariat recognises youth NGOs as one of the categories of groups that may apply for participation as observers.[114] It is important to bear in mind, however, that participation as an observer does not allow such groups to participate directly in negotiations. Rather they can only make general statements in the plenary sessions. Huge numbers of NGOs have attended recent UN climate meetings; for example, 13 500 observers attended the Copenhagen COP in 2009 (Hey and Fourie 2012: 257). The sheer numbers involved have raised issues about how meaningful such participation is in the treaty-making processes (Hey and Fourie 2012: 283). The limited role of NGOs in the treaty-making process reflects a broader lack of democracy at the international level. While treaties may require approval by national parliaments in some states, parliaments generally have limited influence on their content (Crawford 1994: 113–33).[115]

4.5 CONCLUSION

We saw in Chapter 3 that justice principles required to ensure intergenerational justice must operate under an *effectiveness imperative*. We have seen in this chapter that the current international climate change regime falls well short of meeting this *effectiveness imperative*. This is reflected in the failure to agree to binding mitigation targets beyond the linked group of countries who have accepted a second Kyoto commitment period. The failure to have effective mitigation targets makes the UNFCCC and Kyoto technology transfer and funding mechanisms extremely weak, analogous to a house with no foundations.

We have seen that some of the justice principles set out in Chapter 3 are reflected in various provisions of the UNFCCC, but many of the provisions are vague or ambiguous. For example, the notion of 'intergenerational equity' is found in Article 3 of the UNFCCC; however, this fell short of the *core human rights principle* which entails that persons born in the future have equal weight vis-à-vis contemporaries. 'Intergenerational equity' as embodied in the UNFCCC gives no indication as to the weight to be given to future generations.

Similarly there is a soft obligation to promote sustainable development in the UNFCCC which entails taking into account the interests of future generations, but this principle also remains vague. 'Common but differentiated responsibilities' in the global climate regime remains ambiguous, with industrialised countries taking the view that it is based on capacity to pay, whereas developing countries regard this as referring to industrialised countries having caused the problem. In developing a new global instrument under the Durban mandate, the US, in particular, is resistant to 'equity'-based arguments, arguing that these are outside the mandate. However, the UNFCCC Article 3 principles are likely to remain important reference points in the ongoing negotiations.

The relatively robust – albeit vague – justice principles reflected in the UNFCCC contrast with the lack of effective and binding emission targets and other essential elements necessary for an effective global regime in the Kyoto Protocol (including compliance mechanism and funding). This can partly be explained by the domination of particular discourses in the climate change negotiations, particularly the discourse of 'marketisation', which seeks to assimilate climate change governance issues with the market.

This chapter also explored the notions of 'sustainable development' and 'intergenerational equity' in general international law. We found there is no consistent formulation of these concepts which can help in interpreting Article 3 the UNFCCC. Moreover, the indeterminate nature of both concepts means that they cannot give rise to clear binding mitigation obligations independent of the global climate change treaty regime. Both principles, nevertheless, have normative force and have a role in conjunction with other rules of international law, as evidenced in the survey of recent ICJ cases including the *Pulp Mills* case.

Bilateral claims based on state responsibility for transboundary harm were discussed as a possible vehicle for increasing political pressure on states to take effective climate change mitigation action. Importantly, such claims are not a substitute for policy action reflecting the *effectiveness imperative* and justice principles set out in Chapter 3. It was observed that claims in relation to climate change damage face acute difficulties in establishing causation. A further difficulty is the lack of clear thresholds given the limited coverage of the Kyoto Protocol in terms of emission targets. Moreover, it remains unlikely that small island countries threatened by climate change inundation will initiate claims owing to a fear of retaliation.

In relation to procedural justice, the current international system does not contain mechanisms which give explicit representation to future generations through, for example, an international commission for future generations. (This and other comparable mechanisms are discussed further in Chapter 7.) Youth climate NGOs can seek to represent future generations' interests in the ongoing global climate change negotiations. However, under the current UNFCCC rules on NGOs, access to the negotiations remains limited.

In summary, current international law fails to embody effective substantive and procedural justice principles necessary for delivering justice for future generations in relation to climate change. The next chapter seeks to explain this inadequacy. This understanding is a precondition for meaningful recommendations for reform.

NOTES

1. The other regimes are outside the scope of this book. It needs to be borne in mind that international law rules relating to climate change are being promulgated outside the UNFCCC regime by other international organisations *globally* – such as the International Maritime Organization – and *regionally* – for example the European Union's various initiatives, including its Emissions Trading Scheme (http://ec.europa.eu/clima/policies/ets/index_en.htm accessed 19 December 2012). This book limits its scope in this respect to the UNFCCC rules.
2. As of 2 April 2013 the UNFCCC had 195 parties (www.unfccc.int/essential_background/convention/status_of_ratification/items/2631.php accessed 2 April 2013).
3. The Statute of the ICJ provides in Article 38(1) that: 'The Court, whose function is to decide in accordance with international law such disputes as are submitted to it, shall apply: a. international conventions, whether general or particular, establishing rules expressly recognized by the contesting states; b. international custom, as evidence of a general practice accepted as law; c. the general principles of law recognized by civilized nations …' (http://www.icj-cij.org/documents/index.php?p1=4&p2=2&p3=0 accessed 19 December 2012).
4. The *Case of the S.S 'Lotus' (France v Turkey)*, [1927] PCIJ Rep, Series A, No. 9, at 18.
5. In Bodansky et al. (2007: 453–5).
6. Dupuy acknowledges that it is very difficult to say precisely when a rule passes over the 'threshold' from 'soft' law and crystallises into a generally binding rule (Dupuy 2007: 455).
7. [1969] ICJ Rep, 43. As observed by the International Law Association (ILA), there are inherent difficulties in identifying precisely the rules for the creation of customary international law owing to the informal process of its creation, the fact that entails 'controversial questions of the legal theory and ideology' and a lack of systematic treatment of the issue by the ICJ (ILA, London Conference 2000) 3. http://www.ila-hq.org/en/committees/index.cfm/cid/30.
8. *Nicaragua* case, ICJ Reports, 1986, pp 99–100.
9. Article 3(1).
10. The 'Industrialized country leadership principle' found in Article 3 of the UNFCCC is discussed further below at 4.28.
11. The US, Canada, Australia and some other industrialised countries thwarted a push by the EC to include a list of mandatory policies and measures in the Protocol (Sands and Peel 2012: 287).
12. Copenhagen Accord paras 4–5.
13. Copenhagen Accord para 4.
14. Copenhagen Accord para 5.
15. Preamble paragraph 1 Decision 1/CP.17.
16. Decision 1/CP.17 Preamble para 2.
17. FCCC/KP/CMP/2012/L.9 (8 December 2012).
18. FCCC/KP/CMP/2012/L.9 (8 December 2012) 8.
19. The EU made clear that it was willing to increase this to 30 per cent if certain conditions were met. FCCC/KP/CMP/2012/L.9 (8 December 2012) 8.
20. Decision 1/CP.17 para 2, 'Report of the Conference of the Parties on its 17th session, held in Durban from 28 November to 11 December 2011' UN Doc. FCCC/CP/2011/9/Add.1, 15 March 2012.
21. Durban decision 1/CP.17 para 4.
22. Para 7.

23. Decision 1/CP.17 para 8.
24. Para 14, Summary of roundtable on work stream 2, Ad hoc Working group on the Durban platform for enhanced action www.unfccc.int/files/documentation/submissions_from_parties/adp/application/pdf/adp_rt_workstream2_26092012.pdf (accessed 19 December 2012).
25. In 1998 the Global Environment Facility (GEF) of UNDP, UNDP and IBRD was agreed to be the financial mechanism operating on a long-term basis (Sands and Peel 2012: 283).
26. Copenhagen Accord para 8.
27. Decision 3/CP.17.
28. Para 3 of FCCC/CP/2011/Add.1 Annex, 58.
29. Paras 29, 30, ibid. 62.
30. FCCC/CP/2011/Add.1 Annex 63 para 35.
31. Referred to in Lefeber (2012: 328).
32. Statement made on 9 December 2009 by Todd Stern, US delegate to UNFCCC COP15, quoted in Lefeber (2012: 321).
33. Decision -/CP.18 Approaches to address loss and damage associated with climate change impacts in developing countries that are particularly vulnerable to the adverse effects of climate change to enhance adaptive capacity.
34. In Rayfuse and Scott (2012: 290–320).
35. *Earth Negotiations Bulletin*: page 5, volume 12, number 555, Saturday, 8 September 2012: Ecuador proposed a compliance regime linked to the ICJ whereas Brazil proposed the WTO and the Treaty on the Non-Proliferation of Nuclear Weapons as models which the climate change regime could consider.
36. *Case concerning Pulp Mills on the River Uruguay (Argentina v Uruguay)* ICJ Reports 2010, 79, para 281. However, in this case the duty of cooperation was established by the bilateral treaty in question: the 1975 statute.
37. Preamble para 5.
38. Article 4(1)(c).
39. Sands and Peel (2012: 228) argue that the precautionary principle 'has now received sufficiently broad support to allow a strong argument to be made that it reflects a principle of customary law, and that within the context of the European Union it has now achieved customary status' (Sands and Peel 2012:159–64). However, they acknowledge that recent ICJ and WTO panel decisions have refrained from reaching a conclusion that the precautionary principle is part of customary international law. They refer to the *EC – Biotech* case decided by a WTO panel in 2006 (WT/DS291/R, 29 September 2006 para 7.89). The hesitation of these tribunals reflects the fact that a significant number of states reject the proposition that the precautionary principle is binding under customary international law.
40. Bodansky (1993: 501) cites Ronald Dworkin's notion of legal principles contrasted with legal rules: 'rules are applicable in an all-or-nothing fashion' whereas a principle 'states a reason that argues in one direction, but does not necessitate a particular decision'. Principles of law must be taken into account by officials in making decisions. Ronald Dworkin (1977), *Taking Rights Seriously* 24, 26 (Harvard University Press).
41. The ambiguity created in CBDR owing to the differences between its formulation in Article 3 of the UNFCCC and Principle 7 of the Rio Declaration has resulted in conflicting views amongst international lawyers, with some arguing that CBDR rests more on the level of economic development rather than being based on differential responsibility for causing an environmental problem (see Rajamani 2006: 137).
42. An ILA Committee considering legal principles relating to climate change reached the same conclusion (ILA 2010: 11).

43. See also Brunnee and Toope (2010: 153).
44. Atapattu (2009: 42) argues that the one-sided nature of the obligations in the Kyoto Protocol is problematic as it ignores the element of CBDR which refers to the common obligation to protect the environment, as formulated for example in Principle 7 of the Rio Declaration.
45. Statement by China on behalf of Brazil, India, South Africa and China at COP18 made 26 November 2012, Doha, Qatar, www.ccchina.gov.cn/en/NewsInfo.asp?Newsld=34140 (accessed 21 December 2012). See also Statement on behalf of India at the Workshop on Equitable Access to Sustainable Development, 16 May 2012, www.unfccc.int/meetings/boun_may_2012/workshop/6658.php (accessed 19 December 2012): 'since we have agreed that we are not negotiating a new Convention under the Durban platform, the principles of equity will have to apply fully and comprehensively to this process' (4).
46. Preamble, para 20.
47. Decision I COP16 Para 6, FCCC/CP/2010/7/Add.1 (15 March 2011).
48. Preamble, para 2.
49. Article 3, para 4.
50. Principle 3.
51. ICJ Reports 1969, 3, 25.
52. Convention on Biological Diversity, opened for signature 5 June 1992, 1760 UNTS 79 (entered into force 29 December 1993), Article 2.
53. Rio Declaration on Environment and Development, concluded 14 June 1992, Rio de Janeiro, Brazil, Principle 3.
54. Convention Concerning the Protection of the World Cultural and Natural Heritage, opened for signature 16 November 1972, 1037 UNTS 151 (entered into force 17 December 1975).
55. Ibid., Article 4.
56. Bastmeijer assesses the extent to which the Antarctic Treaty System's instruments implement the concept of intergenerational equity using a notion of intergenerational equity built on Brown Weiss and Norton's theories, with a strong emphasis on the need for an ecosystem approach (Bastmeijer 2011).
57. Montreal Protocol for Substances that Deplete the Ozone Layer, opened for signature 16 September 1987, 26 ILM 1550 (entered into force 1 January 1989).
58. Karen Litfin, *Ozone Discourses*: *Science and Politics in Global Environmental Cooperation* (Columbia University Press, New York, 1994).
59. Parliament of Australia Parliamentary Library, *Montreal Protocol on Substances that Deplete the Ozone Layer* (2010) at www.aph.gove.au/library/pubs/Climate Change/governance/international/montreal.htm (last accessed 17 January 2011).
60. 1969 ICJ Reports 1969 3 at [41]–[42]. Some have criticised this requirement; for example, see M.E. Villiger, *Customary International Law and Treaties*, Martinus Nijhoff 1985 (190–202), referred to in Cordonier Segger (2008: 122).
61. See survey by Cordonier Segger (2008: 135), according to which '[n]early every state has some form of environmental law in place which commits it to sustainable development' and several Latin American states incorporate sustainability into their constitutions (136).
62. South Africa's National Environmental Management Act 107 of 1998.
63. *Taralga Landscape Guardians Inc v Minister for Planning* 2007 NSW LEC 59.
64. This follows Vaughan Lowe's idea of sustainable development as a 'meta principle pushing or pulling the boundaries of true primary norms when they threaten to overlap or conflict with each other' (Lowe 1999: 31).
65. *Legality of the Threat or Use of Nuclear Weapons (Request by the United Nations General Assembly for an Advisory Opinion)* International Court of Justice [1996] ICJ Rep 226.

66. Ibid., para 29.
67. Ibid., para 30.
68. *Legality of the Threat or Use of Nuclear Weapons (Request by the United Nations General Assembly for an Advisory Opinion)* International Court of Justice, [1996] ICJ Rep 226 Separate Opinion of Justice Weeramantry, at 383.
69. Ibid., 405.
70. Ibid., 405. He then refers to a number of treaties including the World Heritage Convention (see above note 52), which 'expressly incorporate the principle of protecting the natural environment of future generations, and elevate the concept to the level of binding State obligation'. He also finds support for the principle in academic writings (referring to Edith Brown Weiss's *In Fairness to Future Generations*), 'traditional legal systems across the globe', and the ideals of the United Nations Charter, which 'look forward to social progress and better standards of life for "succeeding generations"'.
71. *Gabcikovo-Nagymaros Project (Hungary/Slovakia)* [1997] ICJ Rep 7.
72. Ibid., para 140.
73. See above note 39.
74. [1997] ICJ Rep 92.
75. *Request for an Examination of the Situation in Accordance with Paragraph 63 of the Court's Judgment of 20 December 1974 in the Nuclear Tests (New Zealand v France)* case, [1995] ICJ Rep 288.
76. In the 1974 *Nuclear Tests* case the ICJ avoided deciding upon the legality of France's atmospheric nuclear testing in Mururoa atoll in the Pacific on the basis that France had made an assurance that future testing would occur underground. *Nuclear Tests Case (New Zealand v France) (Judgment*) [1974] ICJ Rep 457.
77. [1995] ICJ Rep 317 at 341.
78. Ibid.
79. Ibid., 64. Interestingly, in the same case Justice Palmer, in his dissent, also referred to intergenerational equity. He argued that 'this case has to be understood as an environmental case', stating that the 'consequences of these activities need to be carefully analysed and examined unless we are to imperil those to come after us'. He then quoted with approval Edith Brown Weiss's concept of each generation being a trustee of the planet for future generations, arguing that the environmental issues surrounding nuclear testing and nuclear accidents 'demonstrate that States have been unwilling to act as good stewards for or guardians of the environment', suggesting that 'environmental rights ought to be established at the international level and be enforceable there'. Justice Palmer expressed regret that in the case the Court did not take the opportunity to make a contribution to 'one of the critical environmental issues of our time'.
80. Award of the Arbitral Tribunal, 24 May 2005, www.pca-cpa.org/upload/files/BE-NL award corrected 100905.pdf (accessed 16 January 2009). The analysis of the award relies on Cordonier Segger (2008: 124–6).
81. ICJ, CR 2009/17,P57, para 30, quoted in Separate Opinion of Judge Cancado Trindade, [2010] ICJ Rep 135 para 123.
82. (Translation by the author) ICJ, CR 2009/20,P35. French original, quoted in Separate Opinion of Judge Cancado Trindade, [2010] ICJ Rep 135 para 123.
83. *Case concerning Pulp Mills on the River Uruguay* (*Argentina v Uruguay*) judgment, [2010] ICJ Rep 14 para 177.
84. Ibid., para 177.
85. Ibid., para 119.
86. Ibid., para 194.
87. Ibid., para 114.
88. Ibid., para 115.

89. Ibid., para 116.
90. Ibid., para 117.
91. UNESCO Declaration, see p26.
92. Reproduced in Browne Weiss (1989: 293–5).
93. *Case concerning Pulp Mills on the River Uruguay* (*Argentina* v *Uruguay*) judgment, [2010] ICJ Rep 14 para 120.
94. Para 122.
95. Ibid.
96. IACtHR Judgment (merits) of 31 August 2001.
97. Paras 125–30.
98. Ibid., para 149 of the Mayagna judgment, para 128 in Trindade judgment.
99. Ibid., para 130.
100. Ibid., paras 132–8.
101. Ibid., para 138.
102. *Case concerning Pulp Mills on the River Uruguay* (*Argentina v Uruguay*), [2010] ICJ Rep 14 Joint Dissenting Opinion of Judges Al-Khasawneh and Simma.
103. Ibid., para 26.
104. Ibid., para 26.
105. Verbatim Record, Public Sitting of the ICJ, 26 June 2013, in the *Case concerning Whaling in the Antarctic (Australia v Japan; New Zealand intervening)*, Professor Boisson de Charzournes, Agent for Australia (in French: translation by the author) http://www.icj-cij.org/docket/files/148/17390.pdf#view=FitH&pagemode=none&search=%22australia%22 (accessed 27 August 2013), paras 50, 51; p56. The Australian Memorial describes the International Whaling Commission's agenda to conserve whale stocks for future generations reflecting the growing environmental awareness since the 1972 UN Stockholm Conference which involved resolutions calling for the need to preserve nature for the benefit of future generations (Memorial of Australia 9 May 2011, http://www.icj-cij.org/docket/files/148/17382.pdf (accessed 27 August 2013) p60.
106. Stephens relies on Franck (1995).
107. *Case concerning Pulp Mills on the River Uruguay* (*Argentina v Uruguay*) (merits) Judgment of 20 April 2010, [2010] ICJ Rep 14, 203.
108. International Law Commission (ILC) Draft Articles on Responsibility of States for Internationally Wrongful Acts (2001) Report of the ILC on the work of its 53rd session, Official Records of the General Assembly, 56th session Supp No. 10 (A/56/10), chp. IV.E.I.
109. Copenhagen Accord para 5; UNFCCC Decision I/CP. 16 paras 49–50.
110. *Advisory Opinion on the Legality of the Threat of the Use of Nuclear Weapons*, [1996] ICJ Rep 226 para 29.
111. See Strauss (2009).
112. Bilateral claims made have some impact on proposed power plants; see interesting discussion of Micronesia claim against the Czech Republic in Lefeber (2012).
113. UNFCCC Article 7(6); Kyoto Protocol Article 13(8).
114. See NGO contact details form, available at http://unfccc.int/parties_and_observers/ngo/ngo/items/3667.pph (accessed 21 December 2012).
115. Referred to in Hey and Fourie (2012: 283).

5. International human rights law, intergenerational justice and climate change

5.1 INTRODUCTION

Human rights provides a solid basis for an *ethical* obligation on current generations to protect the welfare of future generations in the context of climate change (see Chapter 2). This chapter turns to consider the related but distinct question of whether utilisation of international human rights *law* can assist in meeting the justice principles essential for addressing intergenerational justice set out in Chapter 3.

International human rights law may come into play on a number of levels. Firstly, the breach of human rights obligations which flows from unmitigated climate change may provide further political weight to pressure governments to take necessary mitigation action and strengthen the global climate regime. Couching claims in terms of human rights can give increased political weight to an issue (Hiskes 2009: 7). The recent introduction of human rights discourse into the UNFCCC negotiations reflects this desire for increased political weight. But given North–South conflicts in relation to human rights, this raises the issue as to whether this will exacerbate the existing North–South stand-off in the UNFCCC negotiations, an issue further considered in Chapter 7 below.

Secondly, human rights may provide benchmarks – in the sense of thresholds – on the basis of which mitigation targets may be embodied in a global climate treaty. Human rights would seem attractive here owing to their basis in widely shared values (2.1). But, as we will see, crafting mitigation targets necessarily involves distributional justice issues, such as how to distribute the mitigation burden between current and future generations. Human rights cannot in itself resolve these distributional issues.

Thirdly, international human rights litigation potentially may provide an avenue for representing – albeit indirectly – the interests of future generations. However, international human rights law typically imposes obligations on states towards their own citizens, with those citizens invoking human rights against the state in which they live (Knox 2009: 196). To exert leverage on governments to take strong mitigation action, what is required is, for example, the citizen of a small island state facing inundation to bring a claim against one of the large emitting states such as the US. The structure of international human rights law presents difficult challenges in making such claims. One such challenge is demonstrating that the international human rights instruments invoked impose extraterritorial obligations. Another key challenge is that of causation, given that climate change impacts – such as storm surges – involve a combination of anthropocentric and natural causes which can be difficult to untangle. A further challenge is the issue of standing, which, in relation to future generations, poses particularly acute difficulties as damage has not yet occurred. Recent US cases in, for example, Texas and New Mexico, involving attempts to extend the 'public trust' doctrine to the atmosphere, suggest interesting strategies for combining human rights with international environmental law doctrines. Under the public trust doctrine, trust property is held by the state for the people now and into the future.

This chapter is structured reflecting the three levels set out above. Thus section 5.2 traces the process of linking human rights law and climate change at the international level, in both the UN human rights systems and UNFCCC. This section also identifies which human rights are threatened by climate change. Section 5.3 assesses the value of human rights as 'benchmarks', on the basis of which change mitigation targets may be crafted. Section 5.4 evaluates the potential for human rights litigation to meet the requirements of intergenerational justice. This includes discussion of litigation under the European Court of Human Rights and the US. Section 5.5 draws conclusions.

5.2 LINKING INTERNATIONAL HUMAN RIGHTS LAW AND CLIMATE CHANGE

Linkage between human rights and climate change at the global level began with the 2007 'Male' Declaration on the Human Dimension of Climate Change in which a group of the world's most vulnerable small and developing states, for the first time in an international agreement, stated that climate change had 'clear and immediate implications for the full enjoyment of human rights'.[1] These states were frustrated at the slow pace of progress in the UNFCCC negotiations and wished to focus attention on the victims of climate change (Limon 2010: 546). The 'Male' Declaration led in turn to the Human Rights Council tasking the

Office of the UN High Commissioner for Human Rights (OHCHR) to conduct a study on the effects of climate change on the full enjoyment of human rights. This study concluded that climate change had 'a range of implications for the effective enjoyment of human rights', including 'direct' effects, such as extreme weather events posing a threat to the right to life, but also 'indirect and gradual' effects on human rights, such as increasing stress on health systems.[2]

The OHCHR Report recognised that existing human rights instruments contain rights which would be infringed by the failure of governments to address climate change. These include the right to a standard of living adequate for health and well-being,[3] the right to the highest attainable standard of health,[4] the human right to life[5] and the right to subsistence.[6] These rights would be infringed by governments' failure to respond to climate change, given scientists' predictions of increased exposure to tropical diseases and deaths from extreme weather events (Caney 2009: 69–90).[7]

The OHCHR Report stated that individual claims for breaches of human rights obligations were problematic owing to the difficulty of attributing climate change impacts to individual states.[8] The Report did, however, suggest that members of the international community were subject to an international legal obligation to cooperate to address climate change which flowed from various human rights instruments.[9] In addition, the Report stated that human rights standards should 'inform and strengthen policymaking'[10] and that states had an obligation to satisfy core human rights obligations and to protect particularly vulnerable groups from the harms involved in climate change.[11]

In response to the OHCHR Report, in 2009 the Human Rights Council adopted Resolution 10/4, which stated that 'climate change related impacts have a range of implications, both direct and indirect, for the effective enjoyment of human rights'.[12] It then went on to list those rights most affected by climate change, following the OHCHR Report mentioned above.[13] Resolution 10/4 observed that those whose rights would be most impacted by climate change included groups vulnerable owing to their geography, gender, age, indigenous minority status and disability.[14]

As well as UN human rights bodies reporting on the impact of climate change on human rights, there has been a gradual introduction of human rights language in the UNFCCC process, pushed by a small group of Latin American countries: Argentina, Bolivia and Chile.[15] This has resulted in specific human rights references in the decisions taken at the 16th Conference of the Parties to the UNFCCC (COP16) in Cancun, Mexico in 2010. The Preamble to the Cancun Agreement recognised the need to consult with a wide range of stakeholders, including youth and the disabled, and that 'gender equality and the effective participation of women and indigenous peoples are important for effective action on all aspects of climate change'.[16] The Preamble went on to emphasise that parties should 'in all climate change related actions fully respect human rights'.[17] The Cancun decision on reducing emissions from deforestation noted the UN Declaration on the Rights of Indigenous Peoples.[18]

Since Cancun, however, there seems to have been a move away from human rights language. At the Durban conference of parties to the UNFCCC held in 2011 in South Africa, there were no references to human rights in the Communique of the Conference.[19] A decision on national adaptation plans refers to adaptation taking place through a 'gender-sensitive, participatory and fully transparent approach, taking into consideration vulnerable groups, communities and ecosystems ...' without referring to human rights as such.[20] The decisions of the UNFCCC COP held in Doha in 2012 also failed to refer to human rights.

A similar trend is manifest in the 2012 UN Rio+20 Conference, which reaffirmed the importance of all human rights, including the 'right to development',[21] affirmed the need to 'promote intergenerational dialogue'[22] and emphasised 'that in order to achieve balance among the economic, social and environmental needs of present and future generations, it is necessary to promote harmony with nature'.[23] The conference, however, did not affirm a human right to the environment or a right to climatic stability and the section of the final report on climate change did not use any human rights language at all.[24]

The recent trend away from including human rights language in UN climate change negotiations and documents makes it difficult to assess whether couching climate change mitigation issues in terms of human rights would have a positive impact in terms of increasing the political will to make deep cuts in GHG emissions to protect future generations. The inclusion of references to the human rights of vulnerable groups has the potential to ensure that these groups' particular interests are better taken into account in climate change policy-making. But framing these issues in terms of human needs and welfare under a justice framework, while not pursued in this book, may have an equal chance of success. While the discussion here has been in terms of human rights discourse, a related important issue is whether the linkage to human rights should occur on an institutional basis by introducing UN specific human rights mechanisms which specifically address climate change (considered in Chapter 7 below).

5.3 HUMAN RIGHTS AS BENCHMARKS

Can we derive from human rights suitable 'benchmarks' for meeting the *effectiveness imperative* (EI) essential for delivering justice to future generations? 'Benchmarks' refer to 'moral thresholds' (Caney 2009) or 'line[s] beneath which no one is allowed to sink' (Shue 1996: 18) in relation to which states' action or inaction can be measured (Rajamani 2010: 417). Knox argues that international human rights law imposes a duty on states to cooperate in order to avoid the 'foreseeable destruction of the human rights of substantial numbers of people by, for example, allowing global warming to continue at levels that would inundate small island states' (Knox 2009: 216). A difficulty with this approach is that the benchmark proposed is vague. Indeed, it is arguably no more precise than the *effectiveness imperative* posited in Chapter 3 based on the objective of the UNFCCC of stabilising GHG concentrations in the atmosphere at levels that would 'prevent dangerous anthropogenic interference with the climate system'.[25] As we have seen, defining precisely what the Article 2 objective means and the level of mitigation required to address a projected risk of inundation are both a matter of values (see Chapter 3). A response to this argument could be that while inherently vague, human rights has currency in reflecting widely shared values, and can be the basis for mitigation policies. But how exactly can human rights be a basis for mitigation policies?

Appropriate mitigation targets to be embedded in national or international instruments could be extrapolated from a combination of science, economics and human rights, with human rights being used to provide the minimum benchmarks which should be protected by the mitigation target. Thus one could seek to construct a mitigation target which took as a starting point maximising a 'human right to subsistence' now and into the future.

Indeed, the so-called 'greenhouse development rights' approach developed by Paul Baer and others (Baer et al. 2010) allocates emissions by protecting a minimum level of individual income assessed to be above the poverty line by using the human right to subsistence. This is combined with a notion of *capacity*, to make emission reductions, and *responsibility* for prior emissions of greenhouse pollution (Baer et al. 2010: 222).[26] This example illustrates that to translate a human right to subsistence into concrete emission reduction targets, assumptions would inevitably have to be made, for example, an assumption as to what in material terms is required to live at a decent standard of living. A 'basic human needs' or welfare approach would arguably point in the same direction if meeting 'basic human needs' was equated with meeting a 'right to subsistence'.

Indeed, mitigation targets which are effective in protecting the welfare of future generations could be built upon distributional justice principles with the matrix of justice being human needs, welfare or capacities (Tremmel forthcoming) without relying upon human rights concepts. However, an advantage of utilising human rights in benchmarks is the wide acceptance of human rights within the international community. But human rights as a benchmark can only take one so far. How sharply mitigation is to occur involves a fairness issue in terms of the cost to current generations weighed against benefits to future generations (Caney 2009: 163). Moreover, as discussed in Chapter 3, emission targets should factor in responsibility for historic emissions, capacity to make reductions and development needs. Human rights discourse cannot in itself address these distributional justice issues integral to the design of mitigation targets.

Put differently, human rights cannot resolve conflicts between rights. Climate change policy inevitably raises conflicts between different human rights. Writing in the context of human rights among existing individuals, Rajamani points out that the focus on individual rights gives no guidance as to how such conflicts should be resolved.[27] She observes that there is a conflict between the rights of 500 million poor people in India to gain cheap electricity – required by the right to development – and the rights of inhabitants of low-lying Pacific Islanders threatened with inundation to their culture and islands.[28] Caney argues that in such cases there should be no trade-offs and luxury emissions in wealthy countries should be cut first, as persons in wealthy countries can make such cuts in emissions and still satisfy what is required by core human rights.[29] Thus such conflicts can only be resolved by principles outside human rights.

A similar point can be made in relation to the obligation of sustainable development, which is in conflict with the right of all peoples to freely dispose of their natural wealth and resources (1966 ICESR, Article 1(2)). Sustainable development – defined as development which meets the needs of the present without compromising the ability to meet future needs[30] – implies an obligation towards future generations. However, the right to natural resources implies current generations can use resources now without heed to the impact of this on future generations. These rights cannot of themselves resolve this conflict.

To sum up, human rights have attraction as benchmarks owing to their currency as reflecting widely shared values. But on closer analysis we have seen that human rights as an overall objective requirement offer no more than the current objective of Article 2 of the UNFCCC to avoid dangerous anthropogenic climate change. While beyond the scope of this book, welfare or needs-based approaches would likely point in the same direction. We have also seen that human rights are limited in that they cannot resolve the *distributional* issues involved in crafting mitigation targets, for example, how to balance current and future generations' interests. Distributional justice principles are required for this task. We saw in Chapter 3 that 'subsistence' and 'equality' principles are strong candidates as distributional principles for allocating of GHG emission burdens between current and future generations.

5.4 HUMAN RIGHTS LITIGATION

Can human rights instruments be invoked in litigation on the basis of the potential impact on future generations of climate change, and thus be used as a way of pressuring governments to adopt stronger mitigation policies that deliver justice for future generations? There are a number of limitations to human rights claims regarding climate change impacts on future generations being brought before national or international courts and tribunals, including establishing standing, causation and dealing with extraterritorial harm. A further obstacle is the reluctance of courts to enter into issues considered to be policy and within the purview of governments or the legislature.

Future unborn generations are obviously not in a position to directly invoke breaches of human rights obligations towards them in bringing claims in relation to climate change now. But can children or adults alive now assert claims in relation to harms which are projected to occur in their own future lives or on behalf of future generations, given the high probability of damage suffered in the future sourced by anthropogenic climate change? These issues – in the context of timber licences – rather than climate change, arose before the Philippines Supreme Court in the *Minors Oposa* case, which recognised the right of current claimants to represent future unborn generations where minors asserted that they represented their generation as well as future unborn generations.[31] But this case rested on particular provisions of the Philippines Constitution and it remains unclear as to whether the case will have any wider impact outside the Philippines. Invocation of intergenerational equity in cases in India and Bangladesh has had mixed success (Fitzmaurice 2009: 141).

The *Oposa* case also represents an example of the revival of the common law public trust doctrine. According to this doctrine, natural resources such as rivers are considered to be trust property held by the state for the people – now and into the future – as beneficiaries. This doctrine has recently been revived in Canada and India (Lord et al. 2012: 41, chapters 7 and 19) and in the US. On the face of it, this doctrine would seem to offer potential in overcoming standing obstacles in relation to future generations impacted by climate change in that a court could compel government action to mitigate climate change to the benefit of future generations, provided that courts would accept that the atmosphere is held in trust for the benefit of persons alive now *and in the future*.

In May 2011 there was filing by minors and their guardians in 50 US states of legal actions coordinated by the not-for-profit organisation *Our Children's Trust* invoking the public trust doctrine with the aim of compelling government action to reduce GHG emissions.[32] In the claim brought against the state of California in *Robin Blades v State of California*, for example, the plaintiffs sought a declaration establishing the atmosphere as a public trust resource with a mandatory obligation on the state of California under the public trust doctrine to preserve the atmosphere 'with safe levels of GHGs for this and future generations'.[33] Interestingly, human rights did not play a direct role in the substantive argument although it was mentioned that global warming was a 'fundamental human rights issue for all'[34] and that a delay in curbing emissions would entail a 'host of human rights' and other impacts for Californians.[35] In seeking to establish standing in this case, the plaintiffs – comprising California children and young adults – did not purport to represent future unborn generations – but rather based their interest on the impact which a failure to curb GHG emissions would have on their own future lives. For example, the petition describes the concerns of Robert Blades (aged 14 at the time of the petition) about the impact that climate change will have on his future, noting that he will be 53 years old in 2050.[36]

Similar claims achieved – at least initial – success in the courts of New Mexico in January 2012[37] and in Texas in August 2012. In the latter case a district court in Austin, Texas, rejected the Texas Commission on Environmental Quality's argument that the public trust doctrine was limited to the conservation of water, but held it applied to all natural resources of the state including the air and atmosphere. In reaching this conclusion the court relied on the Constitution of Texas and stated that the public trust doctrine was not simply a common law doctrine.[38] Similar claims have failed in a number of other states in the US where courts have found that there was no basis for an atmospheric trust under state law (Klass 2012b).[39]

An attempt to establish the public trust doctrine in federal law and extend it to the atmosphere, was rejected by the US Supreme Court in *PPL Montana, LLC v Montana*, where it was held that the public trust doctrine remains a matter of state law and its 'contours ... do not depend on the Constitution'.[40] The court went on to explain that even if the public trust doctrine had formed part of federal law at some stage, Congress had plainly displaced any such doctrine, at least in this context, by passage of the Clean Air Act.[41]

While extension of the public trust doctrine to the atmosphere is likely to have mixed success – with cases likely to be appealed – use of the public trust doctrine has the potential to at least put political pressure on governments to take action on climate change in a manner which highlights the impact on future generations.

A significant obstacle to successful human rights litigation to protect future generations relates to causation. In general terms, it is necessary for a person to demonstrate injury or damage to the enjoyment of their human rights and that this damage has been caused by the relevant government (Knox 2009: 191). Even if a court accepted that a person alive today could represent future generations, it is difficult to see how causation could be demonstrated. It would seem impossible to show that the failure by governments to take mitigation action on climate change now caused damage in the future, prior to such damage actually occurring.

Successful litigation on behalf of future generations would face even greater obstacles than those bringing claims now in relation to climate-change-related damage. As we have seen, in the latter cases, causation has been a difficult issue for two reasons. Firstly, climate-change-related damage may have a combination of natural and anthropogenic causes. Damage from storm surges, for example, may be a result of both governments' failure to mitigate climate change resulting in increased sea-level rise but also natural causes. Unravelling these causes in a precise manner remains an acute difficulty in bringing successful litigation.[42] This challenge also arises in relation to human rights litigation.

In *Noel Narvii Tauira and Eighteen Others v France* (1995) residents of French Polynesia claimed the decision of the French government to resume nuclear testing in the South Pacific in 1995 posed a real, substantial and immediate risk to various rights under the European Convention on Human Rights (ECHR), including the right to life.[43] The European Commission of Human Rights rejected the application, stating that only in 'highly exceptional circumstances' could an applicant claim to be a victim of violation of the Convention owing to the 'risk of a future violation'.[44] To fall within such 'exceptional circumstances' the applicant had to 'produce reasonable and convincing evidence of a likelihood that the violation affecting him personally will occur; mere suspicion or conjecture is insufficient in this respect'.[45]

A related limitation is that of making one state liable when GHG emissions are sourced from many countries in various proportions. These difficulties were raised in the case brought by the Inuit people in 2005 before the Inter-American Commission on Human Rights in which they claimed that the US, by failing to curb its GHG emissions, had violated their fundamental human rights.[46] The Commission dismissed the petition without giving reasons but it seems likely that causation was one of the factors involved (Knox 2009: 191).

Where legal action is brought against one of the major GHG-emitting countries, causation issues may not necessarily be an obstacle to successful litigation where a proportionately high contribution to global emissions is involved. The decision of the US Supreme Court in *Massachusetts v EPA*[47] (2007) is interesting in this respect. In this case it was claimed that the US Environmental Protection Agency (EPA) had failed in its responsibility to regulate GHG emissions under the US Clean Air Act. The EPA argued that its potential regulation of new motor vehicles in the US was insignificant in its impact on global GHG emissions, with the US only responsible for a portion of global emissions. The Court rejected this argument, stating that incremental action was significant, pointing out that '[c]onsidering just emissions from the transport sector … the US would still rank as the third-largest emitter of CO_2 in the world … '.[48]

Human rights litigation typically involves a citizen of a state bringing a claim against its own government for a breach of human rights obligations, whereas what is required in the case of climate change harm is to make a claim against several major emitters of GHGs. This raises the issue of human rights claims in relation to so-called 'extraterritorial harms', that is, harm occurring outside the state that is responsible for breaching the particular human rights obligation. This problem would be compounded in relation to potential litigation on behalf of future generations, for example litigation on behalf of a future descendant of a person living in a low-lying island state threatened by climate change against one of the large GHG-emitting countries, for example, a resident of Tuvalu against the US. International law in relation to extraterritorial harms and human rights remains contested, with conflicting views as to whether the International Covenant on Civil and Political Rights (ICCPR) and the International Covenant on Economic, Social and Cultural Rights (ICESCR) impose obligations on the state parties to these treaties not to cause harm to individuals outside the territory of those state parties (Knox 2009: 200–212, Langford et al. 2013: 24).

The ICCPR in Article 2(1) limits state obligations (for example to guarantee the right to life) to 'all individuals within its territory and subject to its jurisdiction'. This has been interpreted by the Human Rights Committee to require a state party to ensure that the rights in the Covenant are given to anyone 'within the effective control of that state party, even if not situated within the territory of the state party'.[49] The ICJ has interpreted Article 2(1) as applying to military control of the state party beyond its own borders (*Israeli Wall* case).[50] However, to date the issue of whether Article 2(1) extends to the victims of transboundary environmental harm has not been considered by any international tribunal (Knox 2009: 203). Moreover, there are conflicting authorities concerning other types of transboundary harm (see Knox 2009: 203–6).

In contrast, the obligations of state parties to the ICCPR with respect to self-determination (Article 1) have no extraterritorial limitation. Article 1(2) includes a right of states to 'freely dispose of their natural wealth and resources' and a requirement not to deprive the people of their 'means of subsistence'. Small island states threatened by sea-level rise could arguably claim that their means of subsistence were being undermined, triggering an obligation on state parties to the ICCPR to extend protection with safeguards, such as environmental impact assessments (Knox 2009: 205, 206). However, to date there has not been a test case applying self-determination obligations outside the decolonisation context.

The ICESCR has no territorial limitation language and, interestingly, Article 2(1) requires state parties to cooperate 'through international assistance and cooperation' in achieving the full right realisation of the rights of the Convention. The Human Rights Committee has consistently argued that the international cooperation obligation in Article 2(1) imposes extraterritorial obligations including in relation to rights to health, water and food. However, the ICJ has taken a contrary view in the *Israeli Wall* case (Knox 2009: 207).

Knox argues that the ICESCR imposes extraterritorial duties even though the exact scope of those duties remains unclear. According to this view, the primary responsibility remains on state parties with jurisdiction over the persons concerned, but states 'in a position to assist other states to meet those obligations are required to do so' (Knox 2009: 207). Knox acknowledges, however, that developed states are strongly opposed to this construction as it is perceived as requiring an open-ended financial commitment (Knox 2009: 208). As noted above, however, the UNFCCC already requires industrialised states to assist developing states in taking adaptation action under the Convention.

An interesting approach which avoids the extraterritorial obstacle is Dinah Shelton's suggestion of a state taking up a claim on behalf of its citizens against major GHG-emitting states (Shelton 2009: 111). Under this approach a state would point to the violations of the state's sovereignty over its natural resources to establish standing to bring a claim. She points to the US Supreme Court's finding of standing in relation to claims brought by various US states in the case of *Massachusetts v EPA*[51] as a model which could bear fruit if attempted at the international level. In that case, the state of Massachusetts established standing on the basis of damage to its coastline from climate change.[52] While Shelton suggests that an international human rights tribunal could be the venue for such a claim, to date interstate human rights claims have not occurred under the relevant ICCPR mechanism; states are reluctant to challenge other states owing to concerns that they in turn will be subject to claims (Hall 2010: 543). Moreover, the *Massachusetts v EPA* approach entailed standing based on concrete injury to the Massachusetts coastline. Establishing causation in relation to damage to future generations *prior to such damage actually occurring* would be problematic even if one could utilise a device such as a 'public trust' which implicitly allowed future generations' interests to be represented.

Finally, litigation on behalf of future generations in relation to climate change harms may face difficulty in that courts are reluctant to interfere with policy issues considered to be within the purview of the government or legislature. While in a broader sense courts implicitly make policy anyway, the courts' involvement in such issues can be perceived as undemocratic and also involve the courts entering into issues in relation to which they do not have the necessary expertise.

In the US this is termed 'the political question doctrine', which prevents the courts from interfering with the other branches of government. This issue arose in 2004 in *American Electric Power v Connecticut*.[53] In this case six electric power companies that ran fossil fuel plants in 20 states in the US brought common law nuisance claims seeking an injunction requiring the power plants to reduce their emissions.[54] The US Supreme Court dismissed the claims for the reason that the EPA had been given the authority to regulate GHG emissions under the Clean Air Act and that this displaced common law nuisance,[55] thus it was 'not for the federal courts to issue their own rules'.[56] However, the court was split on the issue of whether the political question doctrine would prevent common law nuisance claims for GHG emissions.

The European Court of Human Rights has considered a number of cases where breach of the ECHR has been argued in relation to environmental threats.[57] In these cases, however, the Court has been inclined to emphasise due process procedures such as the provision of adequate information or public participation rather than endorsing human rights claims in order to change government policies.[58] This has allowed the Court to avoid conflict with democratically elected governments in fields where it has limited expertise. Moreover, this approach has meshed with the doctrine of 'margin of appreciation', according to which national authorities in tune with local needs are best placed to make regulatory decisions with the court hesitating before interfering with these decisions (*Giacomelli v Italy* 2006).[59] Nevertheless, the Court has found a breach of the Article 2 right to life in a situation where the government failed to impose any planning controls in an area at risk of flooding, leading to citizens' lives being placed at risk (*Kolyadenko v Russia*).[60]

We can see that litigation based on infringements of various human rights in relation to climate change faces serious hurdles, including causation and the need to establish harm or a risk of harm that is not too remote. While establishing standing and causation in relation to climate-change-related harms suffered by persons now is difficult, these difficulties are compounded in relation to harms relating to persons living in the future. In the European human rights system, legal claims on behalf of future generations' projected suffering of climate change damage are precluded by the requirements that claimants show that their specific interests are impacted or a real or imminent threat to life is demonstrated.[61] Courts are likely to remain hesitant in intervening in areas where governments' policy responses remain controversial and outside the courts' expertise (Boyle 2008: 207–9). This is not to deny that climate change litigation conducted on bases other than human rights, for example based on environmental law doctrines,[62] may have success or that the standing rules relating to human rights tribunals may in future be loosened following the example of the US Supreme Court in the *Massachusetts v EPA* case, and perhaps invoking a version of the public trust doctrine.

5.5 CONCLUSION

This chapter has assessed the value of international human rights law for addressing the requirements of intergenerational justice by focusing on three levels. Firstly, the taking up of climate change within the UN human rights system and the use of human rights discourse in the

UNFCCC process. Secondly, an assessment was made of human rights as 'benchmarks' by considering the value of human rights in crafting mitigation targets which protect the welfare of future generations. And thirdly, an assessment was made of human rights litigation.

We have seen that unmitigated climate change threatens a number of human rights well established in UN human rights instruments, including the rights to life, health and subsistence. However, the recent trend away from including human rights language in the UN climate change negotiations and documents makes it difficult to assess whether couching climate change mitigation issues in terms of human rights would have a positive impact in terms of increasing the political will to make deep cuts in GHG emissions to protect future generations.

Human rights are attractive as benchmarks owing to their currency as reflecting widely shared values. However, on closer examination we have seen that human rights as an overall objective offers no more than the current objective of Article 2 of the UNFCCC to avoid dangerous anthropogenic climate change. We have seen that the right to subsistence can constitute the basis for mitigation targets to protect the welfare of future generations, but further assumptions must be made in defining a minimum threshold for subsistence based on needs or welfare. Indeed, welfare- or needs-based approaches would seem to point in the same direction as a human rights approach.[63] Further, we have also seen that human rights are limited in that they cannot resolve the *distributional* issues involved in crafting mitigation targets, for example, how to balance current and future generations' interests. Distributional justice principles are required for this task.

International human rights litigation faces significant challenges, owing to the difficulty of establishing standing for future generations prior to harm actually occurring; difficulty in establishing causation; and difficulty in attributing responsibility in relation to extraterritorial harms. However, one cannot exclude the possibility of the notion of a trust being used in future cases to allow a group to represent future generations. Human rights litigation at the national or international level may potentially help put political pressure on governments to take action in mitigating climate change. However, such litigation is necessarily largely backward-looking and no substitute for effective mitigation policies embodied in a global treaty. Whether an infusion of human rights language and concepts into the UNFCCC negotiations for a global climate treaty would be a help or hindrance is considered in Chapter 7. The latter chapter also considers issues of human-rights-related governance, including whether, for example, a high commissioner for human rights and climate change should be created.

NOTES

1. 'Male' Declaration on the Human Dimension of Global Climate Change, Male, Republic of Maldives, 14 November 2007, Preamble paragraph 12, www.ciel.org/Publications/Male_Declaration_Nov07.pdf (accessed 1 September 2011).
2. Report on the Office of the United Nations High Commissioner for Human Rights on the relationship between climate change and human rights, Annual Report of the United Nations High Commissioner for Human Rights and Reports of the Office of the High Commissioner and the Secretary-General, Human Rights Council, 15 January 2009, UN Doc. A/HRC/10/61, 29.
3. 1948 Universal Declaration of Human Rights (UDHR), Article 25. Adopted on 10 December 1948, G.A. Res. 217 A (III) 3 UN GAOR, UN Doc. A/810, (1948) 71.
4. 1976 International Covenant on Economic, Social and Cultural Rights (ICESCR), Article 12(1) and (2)(b) ICESCR adopted 1966 www2.ohchr.org/english/law/cescr.htm (accessed 16 May 2010) (entered into force 3 January 1976).
5. 1966 International Covenant on Civil and Political Rights (ICCPR), Article 6(1) ICCPR adopted 16 December 1966, 999 UNTS 171 (entered into force 23 March 1976).
6. The rights to an adequate standard of living, health and life are all found in the UDHR (Article 25(1), Article 3) and the ICESR includes both the rights to an adequate standard of living (Article 11) and to health (Article 12).
7. For the health-related impacts of climate change, see Chapter 1.
8. OHCHR Report, (n 2 above) 23.
9. Ibid., 27, 30.
10. Ibid., 26.
11. Ibid., 25.
12. Human Rights Council (UNHRC) Res. 10/4 at 11, UN Doc. A/HRC/10/29 (November 2009) Preamble, paragraph 7.
13. Ibid.
14. Ibid., 564.
15. Rajamani (2010: 400–406).
16. Decision 1/CP.16, 'The Cancun Agreements: Outcome of the Work of the Ad Hoc Working Group on Long-Term Cooperative Action under the Convention', Report on the Conference of the Parties on its 16th Session, held in Cancun, 29 November to 10 December 2010, FCCC/CP and/2010/7/Add.1 (15 March 2011) Preamble, paragraph 7.
17. Ibid., Preamble, para 8.
18. Decision 1/CP.16, 'The Cancun Agreements: Outcome of the Work of the Ad Hoc Working Group on Long-Term Cooperative Action under the Convention' Add.1 paragraph 70 and Appendix 1, 1(c), 12, 26.
19. Report of the Conference of the Parties on its seventeenth session, held in Durban from 28 November to 11 December 2011, FCCC/CP/2011/9/Add.2 (15 March 2012) http://unfccc.int/resource/docs/2011/cop17/eng/09a02.pdf.
20. Decision 5/CP.17 National Adaptation Plans, contained in the Report of the Conference of Parties and work of its 17th session held in Durban 28 November to 11 December 2011, UN Doc. FCCC/CP/2011/9/Add.1 (15 March 2012) 80.
21. Paragraph 8, draft resolution submitted by the President of the UN General Assembly containing the outcome document of the UN Conference on Sustainable Development held Rio de Janeiro 20–22 June 2012, UN Doc. A/66/L.56 (24 July 2012) http://daccess-dds-ny.un.org/doc/UNDOC/LTD/N12/436/88/PDF/N1243688.pdf?Open Element (accessed 29 August 2012).
22. Ibid., para 50.

23. Ibid., para 39.
24. Ibid., paras 190–92.
25. Article 2.
26. For criticism of this approach see Page (2008: 572).
27. Rajamani (2010: 416).
28. Ibid.
29. Caney (2010a: 262–3).
30. *Brundtland Report* (1987) and Rio Declaration on Environment and Development.
31. *Minors Oposa et al. v Secretary of the Environment and Natural Resources Fulgencio Factoran*, GR No. 101083, 30 July 1993, reprinted in (1994) 33 ILM 173.
32. www.ourchildrenstrust.org/ (accessed 13 August 2012).
33. *Robin Blades v State of California*, Complaint of 4 May 2011, www.eenews.net/assets/2011/05/05/document_gw_04.pdf (accessed 13 August 2012), para 8. In February 2012 the petitioners withdrew the petition but without prejudice to refiling the claim (http://ourchildrenstrust.org/state/california accessed 13 August 2012).
34. Ibid., para 35.
35. Ibid., para 10.
36. Ibid., para 13.
37. *State of New Mexico Santa Fe County First Judicial District Court Akilah Sanders-Reed, by and through her parents Carol and John Sanders-Reed, and Wildearth Guardians, Plaintiffs v Susana Martinez, in her official capacity as Governor of New Mexico, and State of New Mexico* No. D-101-CV-2011-01514 (http://ourchildrens trust.org/sites/default/files/Order%20Denying%20Motion%20to%20Dismiss.pdf accessed 5 September 2012).
38. *Angela Bonser-Lain, Karin Ascot v Texas Commission on Environmental Quality*, Cause No. D-1-GN-11-002194, Travis County, Texas, 2 August 2012.
39. For example *Butler v Brewer*, Arizona Court of Appeals, March 2013, 1 CA-CV 12-0347. For a list of these public trust cases see http://web.law.columbia.edu/climate-change/resources/us-climate-change-litigation-chart (accessed 11 September 2013). See also Klass (2012: 1021).
40. *PPL Montana*, U.S., 132S. Ct. 1215, 1235, 182 L. Ed. 2d 77 (2012).
41. Ibid., 15–16. Further litigation on these issues has continued in the US District Court of Columbia. On 27 June 2013 various youth groups announced that they would appeal this District Court's dismissal of their atmospheric trust case in which they sought orders from the court requiring the Obama administration to immediately implement a climate recovery plan involving sharp reductions in GHG emissions. In these cases the youth groups sought to read down the Supreme Court's decision in Montana. They argued that the public trust obligations of the federal government extended to both present and future generations, engaging due process obligations under the 5th and 14th amendments of the US Constitution. They further contended that 'equal protection before the law' was breached by the failure of the federal government to take action as young people and 'posterity' were disproportionately impacted by climate change compared with current generations (see memo to reconsider filed 6/28/12, *Alec L., et al. v Lisa P. Jackson, et al.*, US District Court for the District of Columbia Case No. 1: 11-cv-2235 (RLW) at 21). Interestingly, this legal argument mirrors the notion of equal entitlements or rights discussed in 3.5.1 above.
42. However, attempts have been made to invoke proportionate liability principles – see for example Farber (2008: 401).
43. *Noel Narvii Tauira and Eighteen Others v France* (1995) 83 – B Eur Comm HR 112 315.
44. Ibid., 130–31.
45. Ibid., 131.

46. 'Petition to the Inter-American Commission on Human Rights seeking Relief from Violations Resulting from Global Warming caused by Acts and Emissions of the United States' (December 7 2005) http://www.inuitcircumpolar.com/files/uploads/icc-files/FINALPetitionICC.pdf (accessed 27 October 2011).
47. *Massachusetts v EPA*, 549 U.S. 497, 127 S. Ct. 1438, 1457 (2007).
48. Ibid.
49. General Comment 31, 'The Nature of the General Legal Obligation Imposed on State Parties to the Covenant', U.N. Doc. CCPR/C/21 Rev.1/Add. 13 (26 May 2004) quoted in Knox (2009: 202).
50. Legal Consequences of the Construction of a Wall in the Occupied Palestinian Territory, 2004 ICJ Reports 136, para 127.
51. *Massachusetts v EPA*, 549 U.S. 497, 127 S. Ct. 1438, 1457 (2007).
52. Ibid., at 522–3
53. Ibid., 582.
54. Ibid., 591.
55. *American Electric Power Co. v Connecticut,* 2011 WL 2437011 at 4. Discussed in Michael B. Gerrard and Gregory E. Wannier, chapter 20, 'United States of America' in Lord et al. (2012: 583).
56. The precedential value of the case remains somewhat uncertain. See Gerrard and Wannier (2012: 97) in Lord et al. (2012: 584).
57. See for example *Oneryildiz v Turkey*, 2004 XII 41 Eur. Ct. H. R.20. at 89 and *Budayeva and others v Russia*, decision of 20 March 2008, Appl. No. 15339/02, 21166/02, 20058/02, 11673/02 and 15343/02.
58. For a good analysis of these cases see Pederson (2011: 403–23).
59. *Giacomelli v Italy*, (2006) 5 EHRR 871, at 80. For a discussion of 'margin of appreciation' in relation to climate policies, see Pederson (2011: 421–2).
60. In *Kolyadenko v Russia* (2013) 56 EHRR 2, the European Court of Human Rights affirmed the doctrine of 'margin of appreciation' (para 160) but found that the measures taken by the authorities in Russia were inadequate to discharge Russia's obligations to take positive steps to safeguard the right to life of those within its jurisdiction required under Article 2 of the Convention. The case concerned the release of water from a reservoir which had put at risk the lives of a number of people. The Court found that although the weather conditions which required the release of water were exceptional, the authorities knew that it may be necessary to release the water and the potential consequences should have been foreseen (paras 163–5). The authorities had failed to put in place adequate measures to address this risk (paras 169–73). The case suggests that the 'margin of appreciation' will not apply where the preventive measures taken by a government are totally inadequate to a foreseeable risk. While not yet tested directly by the Court, the logic of this case would seem applicable to climate change adaptation measures where a government took measures totally inadequate to a foreseeable risk, for example from storm surges.
61. See for example Human Rights Act 1998 UK, which provides in section 7 that an alleged unlawful act by a public authority may be challenged for breach of ECHR obligations on the grounds that an individual 'is, or would be, a victim of that act' (Section 7(3)). http://www.legislation.gov.uk/ukpga/1998/42/section/7 (accessed 24 July 2012). In *Kolyadenko v Russia* (2013) 56 EHRR 2, the European Court of Human Rights found that a threat to life was sufficient to fall within Article 2 of the Convention, actual death need not have occurred. However, the Court required clear 'risk to his or her life' (para 151), thus only residents who were at home at the time of flooding from a reservoir were held to have had their lives threatened (paras 151–2). This would suggest that the risk of harm to persons in the future from climate-change-related phenomena such as more extreme storm surges would not be

sufficient to fall within Article 2, but a person caught up in such an event who narrowly escapes death would come within this principle. See also note 60 above.

62. Pederson (2011: 417) suggests that the precautionary principle is one such principle of international environmental law which may well 'prove an inspiration for the Court'.
63. An assessment of 'needs-based' or 'welfare-based' approaches is beyond the scope of this book.

6. Climate change discourses and intergenerational justice[1]

6.1 INTRODUCTION

There is a dramatic disconnect between the powerful ethics and justice-based rationales for strong mitigation action on climate change and the extremely weak international climate change regime described in the previous chapters. It is essential to explain this disconnect before making meaningful proposals to reform the existing international climate change regime so as to reflect the intergenerational justice principles set out in Chapter 3 above.

An assumption of this chapter is that international law reflects power relations and is not neutral (Newell 2008: 515).[2] Existing international law rules on climate change can only be properly understood in their broader political, economic and social context.

A second key assumption is that the multilateral climate change regime reflects government negotiating positions. These positions are impacted by dominant discourses and economic interests, and also societal interests expressed through industry and environment NGOs. Put simply, 'discourses' comprise shared 'meaning of phenomena' (Pettenger 2007: 125) or common understandings (Hajer 1995: 62) or framing of environmental problems (see 6.2 below). The approach of this chapter is to use discourse analysis to explain the weak expression of intergenerational justice in the current global climate change regime.[3] This is supplemented by an analysis of key economic interests which underpin the dominant discourses. It is acknowledged that the interests of the state and the discourses are mutually constitutive, each shaping the other (Wendt 1999: 114). Particularly dominant since the 1980s have been the discourses of 'ecological modernisation', 'industrialism', a 'Promethean discourse', weaker forms of sustainable development and 'climate marketisation' (Dryzek 1997, Stevenson and Dryzek 2012). *Ecological modernisation* refers to the notion that economic growth and environmental protection are compatible provided that environmental harm is internalised through, for example, cap-and-trade systems that ensure that the polluter pays (Hajer 1995: 101). This discourse was a reaction against the discourse of 'industrialism', which entailed maximisation of goods and services, and the 'Promethean' discourse, which assumes technology will always provide answers to environmental problems (Dryzek 2007, in Bodansky et al. (2007): 52). Weaker forms of sustainability correspond to the Brundtland approach to sustainable development which permits the earth's resources to be depleted provided the overall combination of natural and human resources passed on to the next generation is in total comparable with that which was inherited (Bosselmann 2006).[4] 'Climate marketisation' involves market mechanisms being the determining feature of climate governance (Stevenson and Dryzek 2012: 4).

While these dominant discourses provide a useful analytical tool, the form in which they have manifested themselves has varied considerably in relation to individual countries. Thus for example the EU has demonstrated strong elements of *ecological modernisation* (Dryzek 1997: 137–41), reflecting the strong influence of Germany's and the Netherlands' endorsement of this discourse, along with environment and some business groups and EU institutional actors such as the Commission (Wurzel and Connelly 2010: 281). China has demonstrated some elements of ecological modernisation and has even used this term explicitly in its environment policy (Zhang et al. 2007) but this has been constrained by China's heavy reliance on coal-fired power stations (Dai and Diao 2010: chapter 15 in Wurzell and Connelly (2010)).[5]

Over the 1980s and 1990s, the US has been characterised as embodying an economic neo-liberalism discourse and not manifesting ecological modernisation (Dryzek et al. 2003: 174).[6] However, this characterisation of the US is inaccurate in relation to recent developments under the Obama administration since 2009 (Szarka 2012). While the US has continued to resist introduction of a federal cap-and-trade climate policy, other elements of regulation, more typical of the discourse of *ecological modernisation*, have manifest themselves, including regulation by the US Environmental Protection Agency (EPA) under the US Clean Air Act and the use of massive subsidies to promote the renewable energy sector (Szarka 2012). However, to date, structural reform necessary for moving towards a low-carbon economy in both the US and the EU has had limited success owing to the success of vested economic interests in protecting those interests and blocking strong change (below 6.5).

These vested interests have successfully projected their interests as broader societal interests. Gramsci's notion of 'hegemony', extended to the international context, provides a useful analytical tool for explaining how particular vested economic interests have to date successfully resisted the necessary structural change entailed in strong climate change mitigation policies (Phelan et al. 2012: 17). Put simply, for Gramsci 'hegemony' entails the exercise of power not through coercive control but through ideologies reproduced by civil society, the church, academia and the media whereby particular interests are projected as the general societal interest (Gramsci 1971: 181).[7] In the climate change context this has entailed particular sectoral interests, such as coal and oil industries, succeeding in projecting their particular economic interests as coinciding with the general interests of society (Phelan et al. 2012: 8). This has occurred against a backdrop of dominant neo-liberal and consumerist ideologies favouring market-based solutions (Levy and Newell 2005: 52). During the Bush administration, carbon-intensive industries used lobbying and public relations campaigns to successfully oppose any meaningful US participation in legally binding climate mitigation efforts (Vanderheiden 2008: 12).

Closely linked to the dominant discourses mentioned above have been continuing highly influential notions of the primacy of economic growth (Jackson 2009) and the substitutability of natural capital (Phelan et al. 2012: 11). Furthermore, climate change policy has tended to be narrowly framed as an economic issue, with a significant number of economists favouring high discount rates (as discussed above, 1.8). High discount rates entail in effect discounting the interests of future generations and a strong rationale for postponing climate change mitigation action.

In the analysis which follows, the role of intergenerational justice within the dominant discourses is a particular focus. However, as we will see, this cannot be isolated from the intragenerational justice discourse. In the global climate change negotiations, a discourse which emphasises intragenerational equity has in effect squeezed out future generations' interests. Large developing countries, such as China and India, have emphasised economic development and the need to address poverty now combined with a justice or equity discourse which, understandably, places the responsibility for taking the lead on mitigation on industrialised countries. As we saw in the previous chapter, the latter notion of responsibility is reflected in the UNFCCC and current negotiating mandates. This, together with the failure by most industrialised countries to reduce greenhouse gas emissions, has undermined the trust required to forge an agreed climate change regime based on a shared conception of fairness (Roberts and Parks 2007) with serious implications for the fate of future generations. This has occurred in a context where it has become clear that an effective climate regime requires mitigation from at least the larger developing countries (Stern 2008), creating conflict between equity and an effectiveness imperative.

The chapter is structured as follows. The notion of discourses and storylines is explained (6.2), before proceeding to apply discourse analysis to the specific discourses that have manifested themselves in the ongoing UN climate negotiations (6.3). The analysis draws on the work of Stevenson and Dryzek. This is supplemented by my analysis of statements at recent UN climate change meetings. Section 6.4 focuses on the intragenerational justice storyline, which is necessary given the interconnection between intergenerational and intragenerational discourses. Section 6.5 looks at recent literature in political economy to explain how some of the key economic interests have underpinned the dominant discourses.

While it is easy to be gloomy about the prospects of rapidly developing a strong climate mitigation regime, discourses – and the underlying economic and other interests which they reflect – are not set in stone. The potential for a shift in discourses in directions that would ensure intergenerational justice in relation to climate change mitigation is taken up in the conclusion (6.6) and further developed in Chapter 7.

6.2 DISCOURSE ANALYSIS

This chapter relies on the approach to discourse analysis of Dryzek (1997) and Hajer (1995). Central is a notion of a 'discourse' and a 'storyline'. Dryzek (1997: 8) defines a discourse as 'a shared way of apprehending the world' which 'rests on assumptions, judgments, and contentions that provide the basic terms of analysis, debates, agreements, and disagreements'. Applied to environmental politics, Hajer (1995: 44) states that 'discourse analysis primarily aims to understand why a particular understanding of the environmental problem at some point gains dominance and is seen as authoritative, while other understandings are discredited'. Hajer (1995) analyses environmental politics by looking at the way environmental problems are conceptualised. Within this analysis he uses the notion of 'storylines', which he defines as 'narratives on social reality' based upon a 'common understanding' (Hajer 1995: 62). Stevenson and Dryzek (2012) have applied discourse analysis to the global climate change negotiations.[8]

6.3 CLIMATE CHANGE DISCOURSES

Stevenson and Dryzek have applied discourse analysis to the UN climate negotiations, initially focusing on the side event programme of the 2009 UNFCCC Conference of Parties (COP15) held in Copenhagen (Dryzek and Stevenson 2011). They have also applied this analysis to the workshops and meetings held within the Ad Hoc Working Group on Long-Term Cooperative Action (AWG-LCA), which was one of the processes established by the Bali Action Plan agreed at the Bali Conference of Parties (COP13) in 2007 to examine long-term cooperative action by all parties up to and beyond 2012 (Stevenson and Dryzek 2012: 5). I supplement this material with my analysis of national statements made at the UNFCCC 2011 Durban COP, 2012 Doha COP and the Bonn Workshop on Equitable Access to Sustainable Development (EASD) held May 2012.[9] With respect to this supplementary material, only a selection of national statements by some key delegations was used, so the results presented here are necessarily tentative.

Stevenson and Dryzek's categorisation of climate discourses included 'mainstream sustainability', which is a class of discourse sharing the common element that action to address climate change is possible 'within the existing parameters of global economic and political order' (Stevenson and Dryzek 2012: 4), with economic growth needing to be channelled in a greener direction. Within 'mainstream sustainability', this process is portrayed as being achievable by 'climate marketisation' whereby 'all aspects of climate governance can be brought under the logic of the market' (Stevenson and Dryzek 2012: 4). This approach is reflected in support for cap-and-trade schemes and other market-based instruments. 'Mainstream sustainability' contains an assumption that future generations will be provided for within the existing global and political order, with economic growth compensating environmental damage. Thus issues of intergenerational justice are assumed not to arise. Alternatively, the requirements of intergenerational justice are asserted to be met by continued economic growth.

The other key discourse within 'mainstream sustainability' is 'ecological modernisation', which, as mentioned above, emphasises that markets alone are not the answer, rather strong regulation by governments is required including, for example, through targets and timetables and regulation to promote new climate-friendly technologies (Stevenson and Dryzek 2012: 4). But crucially, in *ecological modernisation* discourse environment and the economy are mutually reinforcing and supportive of technological innovation which can solve environmental problems (Hajer 1995: 95–103). Other elements of *ecological modernisation* include 'non-hierarchical bottom-up styles of governance'. But stringent regulation by the state through performance standards, such as feed-in tariffs to promote renewed energy, is also a feature of this discourse (McGee and Taplin 2009: 215–17).[10] As mentioned above, *ecological modernisation* is established in the 1987 Brundtland Report's definition of sustainable development as 'development that meets the needs of the present without compromising the ability of future generations to meet their own needs' (World Commission on Environment and Development 1987: 43).[11] Indeed, '[i]ntergenerational solidarity in dealing with the sustenance base has emerged as an undisputed core principle [of ecological modernisation]' (Mol and Sonnenfeld 2000: 5–7). Thus *ecological modernisation* contains a notion of intergenerational justice reflecting 'weak sustainability', namely an obligation to pass on the earth with a combination of natural and human resources at least comparable with what one has inherited rather than a 'strong' sustainability obligation to preserve the earth's integrity combined with ecological justice (Bosselmann 2006).

The discourse of industrialism has worked hand in hand with what Dryzek (2007: 45–6) describes as the 'Promethean discourse', which as mentioned involves, firstly, the notion that human beings can invent new technologies to overcome any environmental problem and, secondly, the idea that resource scarcity will be addressed by a combination of the market and new inventions. In relation to climate change, the *Promethean discourse* involves the notion that market forces will deliver zero- or low-carbon technologies without the need for strong government regulation (Oreskes and Conway 2010: 257–60).

Stevenson and Dryzek identify 'expansive sustainability' discourse as including 'equitable modernisation', which implies – in addition to decarbonisation – an 'equalising imperative: low carbon development should serve human needs while evening out inequalities between the North and South' (Stevenson and Dryzek 2012: 4). This discourse reflects a version of a global distributive justice which has been reflected in the literature on poverty (Pogge) and climate change (Vanderheiden and Caney).

Further discourses include a 'limits discourse', which sees climate change as a symptom of unsustainable development with the earth hitting ecological limits. There is thus an imperative to shift to economic models not requiring constant growth and environmentally destructive consumption (Stevenson and Dryzek 2012: 4).

A discourse of 'green radicalism' entails the view that climate change reflects an 'unjust economic development model' with carbon markets designed to 'serve the interests of the powerful'. *Green radicalism* suggests that human rights and justice are central to addressing climate change (Stevenson and Dryzek 2012: 5).

While all of the discourses mentioned so far assume the scientific basis of climate change, there has also been present, albeit in the 'background … a persistent discourse of scepticism on the existence of climate change and denial of the need to do much about it' (Dryzek and Stevenson 2011: 1868) (hereafter 'scepticism discourse').

Where does intergenerational justice sit within these discourses? According to *ecological modernisation*, the interests of future generations are assumed to largely look after themselves: so long as the internalisation of environmental damage occurs through the polluter pays mechanism, economic growth can continue with the help of technology fixes and existing governance. Indeed, *ecological modernisation* involves foresight and long-term thinking as part of its essential ingredients (Dryzek 2007: 146). *Climate marketisation* is often linked to the discourse of *industrialism*, according to which the interests of future generations will be taken care of by the market. As we have seen (Chapter 1), the flawed practice of applying high discount rates by many economists – in effect discounting the interests of future generations combined with a belief in economic growth making up for environmental damage – has furnished policy-makers with strong arguments against taking strong mitigation action for climate change now. So *climate marketisation*, in its preference for addressing climate change through the market, combined with the strong influence of conservative economists on climate policy (6.5), works against the interests of future generations. However, ecological modernisation may combine with elements of climate marketisation in the sense of incorporation of market-based policies (such as carbon taxes and cap-and-trade regimes) which, if combined with strong regulation, may address the interests of future generations.

I now turn to examine how the discourses summarised above have manifest themselves in the UN climate negotiations. Stevenson and Dryzek's analysis of the AWG-LCA interventions reveal *mainstream sustainability*, and in particular *ecological modernisation*, being reflected in statements made by the EU and Japan which emphasised the need to facilitate technology development and diffusion, and the benefits of shifting to a low-carbon economy that was consistent with economic growth (Stevenson and Dryzek 2012: 5). There was no specific mention of intergenerational justice in these interventions. However, as mentioned above, the implication of these discourses is that the interests of future generations will be taken care of, given that *mainstream sustainability* includes within its definition a balancing of contemporary and future generations.

The EU statement at the Bonn Workshop on EASD in May 2012 similarly reflected *mainstream sustainability* and *ecological modernisation* while specifically referring to the costs of inaction falling disproportionately on the poor and future generations.[12]

The *climate marketisation* discourse was reflected in statements made by the US in the AWG-LCA which emphasised the importance of private sector financing and intellectual property rights (Stevenson and Dryzek 2012: 6). Again, there was no mention of intergenerational justice, but the implication is that the market will ensure future generations' interests are catered for.

The *Promethean discourse* manifested itself in interventions by, for example, the US, which emphasised the need for 'substantial global investments in transformational technologies'.[13]

Expansive sustainability with an *equitable modernisation* discourse was central to all interventions made by the G77 (which includes all developing countries) and China (Stevenson and Dryzek 2012: 6). The equity principle of common but differentiated responsibilities contained in Article 3 of the UNFCCC[14] was given various interpretations, as: (1) favoured treatment for developing countries given industrialised countries' creation of the problem through historical GHG emissions; (2) 'favoured treatment for the most vulnerable countries' whether they be 'low-lying island states', 'least developed' or 'African countries'; and (3) equity as 'differentiation within Annex 1 Parties' (Stevenson and Dryzek 2012: 6).[15]

Green radicalism found expression in interventions from Bolivia, which coordinates the ALBA block (Stevenson and Dryzek 2012: 14). These interventions emphasised that capitalism was destroying the earth with no recognition of human beings or of 'Mother Earth' (Stevenson and Dryzek 2012: 7). NGOs' interventions also reflected *green radicalism*. However, the influence of the ALBA group[16] and the NGOs on the negotiations was limited (Stevenson and Dryzek 2012: 7).

Green radicalism was also evident in Bolivia's statement at the 2012 EASD workshop. Bolivia emphasised the 'right to development', 'rights of Mother Earth', the 'rights of indigenous people' and 'right of people to overcome poverty'.[17] The 'historical cumulative emissions debt' of industrialised countries was argued to give rise to an obligation on these countries to reduce emissions immediately to zero.[18]

Given the pressure on developing countries to address poverty now, I would not necessarily have expected to see specific mention of intergenerational justice in G77 interventions. However, this assumption was wrong and a number of G77 interventions in the AWG-LCA contained specific references to intergenerational justice. India, for example, stated that any long-term emission reduction goal 'would be meaningless without an agreement on the sharing of the emission rights' and that an equitable distribution of emission rights was required which ensured 'both inter- and intragenerational equity'.[19] To achieve this, India argued that a convergence of all individual emission rights was required 'at a point of time in the foreseeable future'.[20] The Maldives, speaking on behalf of the least developed countries (LDCs), stated that:

> ... one of the key principles of the convention is that the parties should protect the climate system for the benefit of the present and future generations of humankind on the basis of equity, in accordance with their common but differentiated responsibilities and respective capabilities. Accordingly the developed country parties should take the lead in combating climate change.[21]

Interestingly, we can see in this and other statements by developing countries, intergenerational justice being linked to the responsibility of industrialised countries to take the lead in reducing emissions, mirroring Article 3 of the UNFCCC.

The implication is that industrialised countries have *global responsibility* for future generations of *all countries*. Industrialised countries contest this view with, for instance, the US emphasising that each individual country has responsibility for its sustainable development and adaptation – albeit assisted by action under the UNFCCC.[22]

In the AWG-LCA, Bangladesh (for the least developed countries (LDCs)) reflected an *equitable modernisation* discourse in calling for a reorientation of global economic growth patterns with 'focus more on sustainable production and consumption'. Bangladesh also reflected *mainstream sustainability* in calling for 'robust emission goals', necessary 'if we want a safer world for our future generations ... without compromising the prospects of sustained economic growth'.[23] Similarly, Ghana emphasised the need to address equity in terms of 'intra- and intergenerational equity' and 'access to ecological space in the sense of not limiting the sustainable development of developing countries'.[24] Ghana extended this analysis to adaptation, emphasising the need to take into account not only mitigation but the impacts of adaptation now both on current and on future generations.[25] Similarly, in the Bonn Workshop on EASD in May 2012, Bangladesh on behalf of the least developed countries (LDCs) emphasised sustainable development 'not just for today but for generations to come'.[26]

Mainstream sustainability was also evident in the statement on behalf of the Association of Small Island States (AOSIS) at the Bonn workshop on EASD in 2012, which emphasised that '[r]educing emissions and achieving sustainable development are compatible' with 'clean energy' providing an 'alternative path to development'.[27]

The discourse of *expansive sustainability* was reflected in the interventions of a number of developing countries at the AWG-LCA including, for example, China and the Philippines. China emphasised the principle of equity as being central to the long-term vision for the Convention, which they interpreted as requiring convergence to equal per capita emissions over time, but with developing countries given leeway to modernise and develop before reducing emissions.[28] China emphasised that 'poverty eradication is the number 1 and overriding priority for developing parties', with these countries needing to develop so they were strong enough to tackle climate change.[29] The Philippines spoke in a similar vein, arguing that action was needed now to allow the country 'to survive and to pursue its sustainable development goals, to eradicate poverty' and to avoid 'compromising the needs of our future generations'.[30]

Expansive sustainability was also reflected in the statement by India at the 2012 EASD workshop, who emphasised the right to development and the 'imperative of poverty eradication' as key elements of 'equity' which were central to implementing the Durban platform.[31]

6.4 THE INTRAGENERATIONAL JUSTICE STORYLINE

We have seen in the statements of a number of developing countries *expansive sustainability* discourses which prioritise addressing pressing poverty concerns now, while some statements also explicitly recognise the need to ensure intergenerational justice. Developing countries have tended to explicitly or implicitly place the *responsibility* for ensuring intergenerational justice on industrialised countries by arguing that future generations' needs will only be met if industrialised countries take the lead in reducing emissions and transfer funds and technology to allow developing countries to respond to climate change impacts.

On the other hand, as we have seen, industrialised countries have reflected *mainstream sustainability* or *ecological modernisation* discourses which implicitly involve a balancing of future generations' interests. It is worth recalling that the Brundtland concept of sustainable development contains within it a strong notion of social justice: sustainable development depends on addressing the needs of the poor as well as the needs of future generations (World Commission on Environment and Development 1987: 43). Indeed, the Brundtland notion of sustainable development entails the needs of future generations constraining development. Thus the Brundtland notion of sustainable development was defined as 'development that meets the needs of the present generation without compromising the ability of future generations to meet their own needs' (World Commission on Environment and Development 1987: 43).

This notion has embedded in it a sufficiency notion of justice: each generation has no obligation to pass on to their successors the same or an improved level of well-being; the obligation is limited to providing to the next generation enough so as not to compromise 'the ability of their successors to lead decent lives' (Page 2006: 91). However, sustainable development at a political level has been attractive to many countries because of its emphasis on economic development, with the social justice dimension receiving less attention. As eloquently put by Dryzek (2007: 57), 'the rich eventually forgot (or never really accepted) the global equity central to Brundtland'.[32]

While Article 3 of the UNFCCC includes both inter- and intragenerational equity (see Chapter 4), arguably North–South tensions reflecting intragenerational injustices have squeezed out intergenerational justice concerns. Thus, in the UN climate meetings in Copenhagen, Cancun, Durban and Doha, Pacific Island countries' calls for deep cuts in GHG emissions to ensure the future survival of their nations have been ignored.[33]

The emphasis we have seen in developing countries' interventions on industrialised countries acting first to make deep cuts in their greenhouse gas emissions – given their historic responsibility in creating the problem and their greater capacity to respond – has been a feature of the UN climate negotiations from the outset (Bodansky 2010: 3). Put in the language of discourse analysis: developing countries have been using a strong intragenerational equity storyline which puts addressing poverty now as a top priority and implicitly squeezes out the interests of future generations. This is understandable. As put by Ross Garnaut (2008b: 2):

> It might make perfect sense for a rich country like Australia to sacrifice some current income for the benefit of future generations. It will not seem quite as simple a matter for a poor country, with most of its people in abject poverty, that needs rather strong economic growth now to get people out of poverty and give people the luxury of thinking about environmental values and the welfare of future generations.

However, given that China has now surpassed the US as the largest producer of greenhouse gas emissions, and that developing countries will soon overtake industrialised countries as a source of greenhouse gas emissions, it has become clear that without significant reductions in emissions by at least the larger developing countries, it will be impossible to reduce greenhouse gas emissions sufficiently quickly to address this problem – thus creating tension between equity and effectiveness.[34]

Are the storylines of intergenerational and intragenerational equity in conflict in relation to climate change? Certainly it would seem that in developing countries, given their weak social security systems, a rapid economic restructuring to reduce greenhouse gas emissions by a shift from coal to renewables and nuclear energy would result in at least a short-term increase in poverty for sections of society. On the other hand, the Intergovernmental Panel on Climate Change (IPCC 2007a: 706–7) has argued that implementing sustainable development in a way that significantly reduces greenhouse gas emissions can be done by developing countries such as China in a manner which does not involve a slowing in economic development.

There are certainly overlaps in considering justice for future generations and justice between generations alive today. While the discourse analysis of Hajer (1995) assumes that interests and power cannot be separated from the modes of communication, the interests of future generations and even younger people alive today are at a considerable disadvantage in the policy-making process, an issue further addressed in Chapter 7.

6.5 INTERESTS AND DISCOURSES

In applying discourse analysis it is essential to focus on the underlying powerful economic interests that underpin public policy decisions and shape discourses on climate change. In a case study of Norwegian climate change policy-making, Tellmann (2012) demonstrates that while discourses constrain the range of policy choices, interest politics ultimately determines which policies are actually implemented.[35]

In addition, a focus on national statements made by diplomats can imply taking at face value positions adopted by countries which – while dressed in the language of fairness – in fact reflect *unstated* economic or strategic interests, linked, for example, to a concern for competitiveness (Winkler and Beaumont 2010: 643).[36]

Newell and Paterson (2010) trace the shift from the 1970s – when the key corporations in the global economy were oil companies and manufacturers – to the 2000s – when the key large corporations were in finance and information technology (Newell and Paterson 2010: 20). In the 1970s through to the late 1980s, oil and other companies concerned at how climate mitigation could impact their businesses, initiated a concerted campaign to fund climate sceptics in the US and elsewhere (Newell and Paterson 2010: 37, Vanderheiden 2008). The message was that either climate change was not happening or that, if it was, it was not human-made and action to reduce emissions would cripple the economy (Newell and Paterson 2010: 38). A small group of scientists in the US from the 1980s through to the 1990s worked to create an impression that the science on climate change was uncertain to deliberately create an impression that all aspects of climate change science were uncertain, including the well-established link between increased CO_2 emissions and climate change (Oreskes and Conway 2010: 34). Their campaign was assisted by a media which persisted in giving climate sceptics the same coverage as mainstream science in spite of only a tiny proportion of scientists supporting sceptics' views (Oreskes and Conway 2010: 214).[37]

Implicit in this sceptic or denial view was that intergenerational justice was simply not an issue: there was no problem to be addressed in the first place. Alternatively, the implication was that climate change was natural, with no anthropogenic dimension, so resources should not be wasted on mitigation but could be legitimately directed at adaptation and that, in any event, human migration would be an acceptable solution (Oreskes and Conway 2010: 180).

The 1980s also saw the spread of economic neo-liberalism, which in climate policy equated with a reluctance to regulate, rationalised by the use of government–business partnerships and a preference for market instruments such as cap and trade (Newell and Paterson 2010: 23–5).

The 1980s also saw *ecological modernisation* come to the fore with a preference for economic instruments to address climate change. Ecological modernisation was particularly influential in Germany and the Netherlands, but then became manifest in EU policy and even outside Europe, for example to some extent in China (Wurzel and Connelly 2010: 281–2).

By the 1990s there was a rise in financiers and service industries who began to see climate change policy as an opportunity, rather than a threat, and a chance to make money as carbon markets were developed; insurance industries began to incorporate climate-related costs in their business strategies (Newell and Paterson 2010: 42). Thus the discourse of *ecological modernisation* came to be underpinned by emergent economic interests. The EU's leadership in global climate negotiations was underpinned by the incorporation of *ecological modernisation* in a number of EU states (Wurzel and Connelly 2010: 280–83). This included incorporation of this discourse in Germany's industry policy with significant employment in renewable energy strengthening pressure for regulation (Janicke 2010: 129–46).[38] However, Germany's coal and automobile industries continued to impose constraints on German governments moving towards stronger regulation (Wurzel and Connelly 2010: 281).

The weak global climate change regime has reflected the continuing dominating influence of fossil fuel, oil and related interests in national government positions. This has included a disproportionate influence of fossil fuel and related high GHG-dependent industries on government policy through well-developed lobbying networks.[39] It has also included the 'discursive' dimension: the societal commitment to carbon-dependent economic growth (Phelan et al. 2012: 17).

Japan's position in the Kyoto Protocol negotiations demonstrated the interplay of particular interests and institutional factors, as well as discourses. After the oil shock of the 1970s, Japan's industry became a world leader in efficiency (Takao 2012: 777). However, Japan did not introduce in the 1980s or 1990s strong regulation in the form of a cap-and-trade scheme or carbon taxes advocated by its Environmental Agency owing to a powerful business coalition (Oshitani 2006).[40]

There were two key reasons for this: firstly, the framing of climate policy as industrial policy and, secondly, the influence of vested interests. From the 1980s climate policy tended to be framed as industrial policy, with Japan's powerful Ministry of International Trade and Industry (MITI) regarding technology as the solution to climate policy (Takao 2012: 777). MITI's Agency of Natural Resources and Energy worked closely with 'the coal, electricity, gas, nuclear and petroleum industries' in managing Japan's energy policy after the two oil crises (Uchihashi 1995: 163–70).[41] However, Japan's Ministry of Foreign Affairs and conservative politicians had deep concerns that Japan enhance its international reputation by playing a leadership role in the Kyoto Protocol negotiations (Takao 2012: 780).

Since April 2012 Japan has a carbon tax set at a relatively low level which came into force on 1 October 2012 and a cap-and-trade scheme at the local/city level covering 20 million people (Climate Commission 2012: 55). Renewable energy is fostered by subsidy and tax systems and fuel efficiency targets have been introduced for automobiles (Climate Commission 2012: 55). Since the 2011 earthquake and nuclear crisis, Japan's long-term energy plans are being reconsidered (O'Sullivan et al. 2012). The replacement of nuclear energy with fossil-fuel-based energy would increase Japan's emissions, at least in the near term (Norton Rose Group 2011).

The 2000s saw a softening of neo-liberalism and in the US the emergence of sectors, such as the insurance industry, which favoured stronger regulation on climate change (Newell 2008: 527). The creation of carbon markets in the EU – since 2005 – and elsewhere led to the emergence of interest groups that counteract vested interests resisting climate change mitigation (Bernstein et al. 2010: 170).

US climate policy-making remained under the Obama administration (from 2009) constrained by a liberal economic framework. At the time of writing the US continued not to have a federal-level cap-and-trade system. However, the Obama administration in 2009–10 launched a number of initiatives aimed at reducing GHG emissions and fostering a shift towards renewable technologies which contain elements of an *ecological modernisation* discourse. These included expenditure of USD167.47 billion in the Obama administration's first two years in office on various programmes designed to foster new renewable energy-related technologies (MacNeil and Paterson 2012: 239). During 2009–10 best available technology standards were prescribed for new energy facilities in the US using Environmental Protection Agency (EPA) rule-making under the federal Clean Air Act (MacNeil and Paterson 2012: 241). Climate change regulation at the state level in the US accelerated in this period and as of 2011 there were 34 states within the US which had climate-change-related arrangements (MacNeil and Paterson 2012: 240).

It remains to be seen whether these initiatives result in US climate policy shifting in the direction of sharp reductions in GHG emissions. President Obama in his second inaugural address in January 2013 promised action on climate change, stating that inaction would 'betray our children and future generations' and that 'none can avoid the devastating impact of crippling drought and more powerful storms'.[42] He called for a massive shift from fossil fuels to renewables. In June 2013 President Obama released a 'Climate Action Plan' which included a direction to the EPA to complete carbon pollution standards for both new and existing power plants, a 30 per cent increase in funding for clean energy technology and further fuel economy standards for vehicles for the years 2014–18.[43] It remains uncertain whether increased expenditure on renewable-related programmes – over time – counteracts 'the entrenched power in Congress of coal interests' opposing climate mitigation and influences in a positive direction efforts for a global treaty (MacNeil and Paterson 2012: 242). Moreover, the US technology policy approach may not necessarily be an effective blueprint for a global climate treaty as it may be undermined by an 'imperative to preserve economic competitiveness' rather than an environment objective (MacNeil and Paterson 2012: 243).[44]

6.6 CONCLUSION

By analysing national statements made in the UN climate negotiations, we can see that industrialised countries reflected discourses of *ecological modernisation* and *mainstream sustainability* which implicitly incorporated intergenerational justice. But industrialised country rhetoric has not been matched by action, with industrialised countries – with a few exceptions – failing to adequately reduce GHG emissions.[45]

Developing countries' statements reflected discourses of *expansive sustainability*, *equitable modernisation* and a storyline of intergenerational justice. Thus developing countries attributed to industrialised countries a *global responsibility* for ensuring protection of future generations from climate change, owing to the latter's historic emissions. This discourse provided a basis for developing countries resisting emission reductions in a context where it has become clear that at least the larger developing countries will need to take on significant emission reductions for any global climate regime to be effective.

The disconnect between the most dominant discourses of mainstream sustainability and ecological modernisation – which prima facie incorporate intergenerational justice – and the failure of governments to reduce GHG emissions is partly explained by the powerful economic interests which support the existing global economic system with its strong reliance on fossil fuels. Important also is the relative power of economic discourses such as *market liberalism* and *climate marketisation*, which have favoured a soft or non-regulatory approach to climate change. While US responses have continued within the parameters of an economic liberalism discourse, under the Obama administration regulation has proceeded at the state level, and even at the federal level through EPA regulation under the Clean Air Act. However, it remains uncertain whether this will result in the significant structural change required to move to a low-carbon economy.

Also crucial has been the framing of climate policy as industry policy rather than environment policy in, for example, the US, Japan and Norway. While subsidies for renewables were consistent with this approach, a federal cap and trade has been off limits in these countries owing to trade competitiveness concerns. (Although Japan introduced a modest carbon tax in 2012 and an emissions trading scheme at the local/city level, as mentioned above.) Strong trade and industry ministries have taken up these industry competitiveness concerns to trump environment ministries. Climate policy was also merged with industry policy in Germany – however in a different context, where environmental concerns had greater leverage partly owing to the success of the Green Party (Wurzel and Connelly 2010: 281).

A significant source of the intense North–South conflict found in the climate change negotiations is the bitterness felt by developing countries in response to their treatment in relation to trade and debt issues – with weaker states reciprocating in relation to issues where they have more leverage – including climate change (Roberts and Parks 2007: 6).

The extent to which prevailing discourses are capable of change is contentious. Giving greater prominence to ethics and justice discourses in climate change policy – and making explicit the ethical assumptions in economic analysis – would be of great value in developing a climate regime policy that ensures justice for future generations. Whether and how this could be brought about is one of the themes taken up in the next chapter.

NOTES

1. This chapter relies on Peter Lawrence, 'Justice and Future Generations: Environment Discourses, International Law and Climate Change' in: Brad Jessup and Kim Rubenstein's (eds) *Environmental Discourses in International and Public Law* (Cambridge University Press, 2012).
2. As stated by Armstrong et al. (2012: 302): 'Some would argue that ... [the international legal system] systematically promotes the interests and values of the wealthier and more powerful states, especially the West.' These authors point to particular areas of international law which favour Western interests, such as the rules relating to intellectual property rights, rules relating to expropriation, as well as many international organisations including the UN Security Council, World Bank and International Monetary Fund (IMF) being 'structured in ways that favour the major Western powers'. They also note more 'subtle inequities' such as the better access to legal advice which wealthier nations can draw upon when disputes occur (Armstrong et al. 2012: 303).
3. While this chapter relies on discourse analysis and literature which sheds light on underlying economic interests to explain the weak embodiment of intergenerational justice in the global climate regime, it is acknowledged that this is only a partial explanation: a more complete explanation would have to rely on social psychology which seeks to explain human beings' motivational difficulties with behaviour aimed at long-term objectives (see references in Gillespie (1997: 117–18). This latter field is beyond the scope of this book.
4. Referred to in Richardson and Wood (2006: 129, 150–55). (Richardson BJ and Woods S (eds), *Environmental Law for Sustainability: A Reader* (Hart Publishing 2006)).
5. Referred to in Wurzel and Connelly (2010).
6. Referred to in Szarka (2012: 92).
7. Levy and Newell (2005: 49–50).
8. See also Dryzek and Stevenson (2011).
9. My analysis of national statements which follows was made possible by transcripts of interventions made in the AWG-LCA meeting and workshops held 31 March to 4 April 2008 Bangkok, Thailand, kindly provided to me by Hayley Stevenson. Hayley Stevenson made these transcripts based on the webcasts of these meetings which *are* available on the website of the UNFCCC at http://unfccc.meta-fusion.com/kongresse/AWG_08/templ/ovw_str_awg_all_sessions.php?id_kongressmain=43.
10. McGee and Taplin rely upon Mol and Sonnenfeld, Janicke and Huber.
11. Hajer (1995: 101).

12. Presentation of the European Union at the Workshop on EASD, 16 May 2012, Bonn http://unfccc.int/files/bodies/awg-lca/application/pdf/20120516_eu_0924.pdf (accessed 4 November 2012).
13. 4.50pm AWG-LCA 1st session 31 March to 4 April 2008, 2 April 2008 Informal Plenary, United States.
14. The text of Article 3 of the UNFCCC is set out in 4.2.7 above.
15. See 4.2.7 above for an analysis of CBDR in the global climate regime.
16. On ALBA see Hayley Stevenson (*Representing Green Radicalism*, forthcoming 2013).
17. Presentation of Bolivia at the workshop on EASD, 16 May 2012, Bonn www.unfccc.int/meetings/bonn_may_2012/workshop/6658.php(20120515_bolivia_2340.pdf (accessed 15 January 2013), 5.
18. Ibid., 6.
19. AWG-LCA 1st session 31 March to 4 April 2008, 2nd meeting 1 April 2008 # 17 India 3.09pm.
20. Ibid.
21. AWG-LCA 1st session: 31 March to 4 April 2008, Bangkok, Thailand, 3rd meeting 2 April 2008, 2.27pm.
22. AWG-LCA 1st session: 31 March to 4 April 2008, Bangkok, Thailand, 4th meeting 2 April 2008, United States, 5.36pm.
23. Ibid., 9.24am.
24. AWG-LCA 4th Session, Poznan, 2 December 2008 workshop: Shared Vision for Long-Term Cooperative Action, 5.25pm.
25. Ibid.
26. Presentation of Bangladesh on behalf of LDCs at the workshop on EASD, 16 May 2012, Bonn http://unfccc.int/files/bodies/awg-lca/application/pdf/20120516_bangladesh_ldc_0912.pdf (accessed 4 November 2012).
27. Presentation of AOSIS to the workshop on EASD, Bonn, 16 May 2012 http://unfccc.int/files/bodies/awg-lca/application/pdf/20120516 (accessed 9 January 2013).
28. AWG-LCA 4th session, Poznan, 2 December 2008, workshop: Shared Vision for Long-Term Cooperative Action, China, 14.34pm.
29. Ibid.
30. AWG-LCA 4th session, Poznan, 2 December 2008, workshop: Shared Vision for Long-Term Cooperative Action, Philippines, 3.20pm.
31. Presentation of India to the Workshop on EASD, Bonn 16 May 2012 www.unfccc.int/meeting/bonn_may2012/workshop/6658.php (accessed 9 January 2013, p2).
32. Dryzek refers to James Meadowcraft (2000: 379).
33. See for example 'Pacific countries disappointed with climate talks', Australia Network News, 7 December 2012 www.abc.net.au/news/2012-12-07/an-pacific-countries-disappointed-with-doha-climate-talks/4414308 (accessed 15 January 2013).
34. Scholtz (2009: 167).
35. Tellmann (2012) demonstrates how Norway from 1989–2008 went from a 'tax discourse' derived from economic rationalism to a 'quota discourse' based on the notion of emissions trading. However, the latter was only weakly implemented. Norway's climate policy remains strongly influenced by an expansion of oil production from the 1990s accompanied by a strong belief in technology solutions such as carbon capture and storage, allowing a move to a low-emission society. However, thus far new technologies have not delivered in terms of emission reductions (Tellmann 2012: 746).
36. In Stevenson and Dryzek (2012: 17). (Winkler H and Beaumont J, (2010) 'Fair and Effective Multilateralism in the Post-Copenhagen Climate Negotiations,' 10(6) *Climate Policy* 638).

37. See 1.1 above.
38. Referred to in Wurzel and Connelly (2010).
39. In the context of Australia see Pearse (2007, 2009).
40. Referred to in Takao (2012: 780).
41. Referred to in Takao (2012: 780). (Uchihashi K, *Kyosei no daichi atarashii Keizai ga hajimaru [The Great Land of Cohabitation]* (Iwanami Shoten 1995)).
42. Suzanne Goldenberg, 'Climate change moved to forefront in Obama's second inaugural address' www.guardian.co.uk/world/2013/jan/21/climate-change-obama-inaugural-address (accessed 12 February 2013).
43. The President's Climate Action Plan, Executive Office of the President, United States, June 2013, http://www.whitehouse.gov/sites/default/files/image/president27s climateactionplan.pdf (accessed 27 August 2013), 6, 7, 8.
44. It is important to note that for effective technology development and diffusion to take place, both 'technology push' in the form of investment and subsidies for technology development are required as well as 'market pull' in the form of government regulation which ensures that there is a market for climate-change-related technologies which are developed. See Lawrence (2008: 43) and references cited therein.
45. See above 1.10.

PART III

The way forward and conclusion

7. The way forward – incorporating intergenerational justice principles into international climate law

7.1 INTRODUCTION

This book has proposed a suite of justice principles that need to be reflected in an international climate regime for that regime to ensure justice for future generations. These principles are respect for *core human rights required to ensure respect for human dignity*, *responsibility for harm* and *capacity to pay*, operating in conjunction with core distributional justice principles of *equality* and *subsistence* (Chapter 3). Whereas previous chapters have examined the content of these principles, this chapter examines how such principles should be incorporated in international law.

There is a vast literature on proposals for a global climate change agreement and methods for distributing mitigation burdens.[1] The focus here is on the role of international law and intergenerational justice in relation to such proposals. 'Effectiveness', as explained in Chapter 3, is of crucial importance. A global climate agreement may respect *intra-generational* justice in the distribution of emissions but fail to meet the requirements of *intergenerational* justice if emissions are reduced too slowly, or not at all, harming the interests of future generations.

We have seen in the previous chapter that the weak embodiment of intergenerational justice in current international law rules on climate change reflects dominant discourses of market liberalism and weak sustainability underpinned by economic interests resistant to the structural change necessary to move towards a low-carbon global economy.

This contextual dimension is vital in that meaningful reform proposals must take place within political-economic constraints. Proposals consistent with the dominant discourses have a greater chance of being accepted. But recall that an attraction of the neo-Gramscian framework is its dynamism: discourses are not set in stone and may change over time. We saw in Chapter 6 that currently, industrialised countries reflected intergenerational justice in their discourse but this was not matched by mitigation action. On the other hand, developing countries placed global responsibility for ensuring justice for future generations on developed countries. It flows from this that for intergenerational justice to be implemented in international law, strategies must be found to effectively combat the dominant economic interests resisting change. Political leadership is vital for this to occur. In addition, a notion of global responsibility must be fostered (Vanderheiden 2011).

Some argue that a shift to a low-carbon economy is possible by reforming the existing capitalist system (Newell and Paterson 2010), whereas others argue that a new economic model is required which ensures sustainability without growth (Jackson 2009). While there is controversy as to whether the first option will be sufficient to deliver sufficiently deep cuts in greenhouse gas emissions, there is a compelling argument that as a first step this path be adopted urgently. Put differently, there is an urgent need to incorporate the cost of environmental damage in the pricing of goods and services. This prescription entails elements of *ecological modernisation*, as described in Chapter 6. But more than this is required. We need a new form of ecological macro-economics for sustainability which entails a move away from GDP growth as an objective and towards 'prosperity' defined in a broader non-materialistic manner (Jackson 2009: 176). Achieving this will require short- and long-term emission reduction targets embedded in an effective global treaty, with a strong compliance mechanism, a price on carbon (carbon tax or cap-and-trade systems), an effective funding mechanism to assist developing countries in shifting to a low-carbon sustainable development and a massive increase in technology development and transfer to help ensure the rapid uptake of renewable energy technologies (Lawrence 2008: 43–8).

Secondly, there needs to be a shift in the dominant discourse of developing countries towards a notion of proportional responsibility, reflecting principle II argued for above (3.5.6). This chapter (following Roberts and Parks 2007) argues that such a shift in the discourse of developing countries will only occur if *mistrust* between the industrialised and developing countries in – and outside – the climate negotiations is addressed. This requires industrialised countries meeting their commitment under the UNFCCC to take the lead in reducing GHG emissions. However, this also requires confidence-building action outside the climate negotiations in relation to, for example, agricultural subsidies and intellectual property rights rules (Templeman 1998) which harm the development prospects of developing countries (Roberts and Parks 2007: 23).

This chapter augments the justice principles argued for in Chapter 3 by 'feasibility constraints', reflecting an assumption that certain political constraints must be complied with for fairness principles to stand a chance of being implemented in the real world. This chapter does not aim to provide a comprehensive review of proposals for distributing mitigation burdens and a global climate change agreement, or to propose a particular blueprint for implementing intergenerational justice. Rather, my aim is to show that it is practically possible to incorporate justice principles in an agreement that is politically feasible, albeit dependent upon strong political leadership.

The chapter is structured as follows. Section 7.2 analyses the issue of whether common justice principles must be agreed upon as a precondition for negotiating an effective global climate regime. Related to this is the issue of whether a common understanding of justice concerning the distribution of mitigation burdens is required or outcomes reflecting a compromise in justice principles is sufficient. Section 7.3 examines 'legal form' issues, addressing whether a legally binding treaty is essential for delivering intergenerational justice or whether other options, such as elaboration of the Copenhagen pledge and review model, are more desirable. Section 7.3.1 examines the issue of whether the consensus model of treaty-making constrains the likelihood of negotiating a treaty which respects long-term interests. Section 7.4 introduces 'feasibility' criteria, which, inter alia, reflect requirements of participation and trade competitiveness concerns. In section 7.5 'feasibility' criteria are combined with the *effectiveness imperative* and justice principles set out above (Chapter 3) to analyse national proposals made in the ongoing negotiations under the Durban platform. Given that the national proposals only partially reflect the Justice and Implementation Principles set out in Chapter 3, I discuss (in 7.5.1) how these principles should be implemented in a global climate treaty, and how the Durban platform would need to be amended to achieve this end. Section 7.6 makes recommendations as to how the requirements of procedural justice in relation to future generations can be incorporated in international law-making on climate change. A particular focus is the possibility of expanding the participation rights of youth justice NGOs on the basis that they have interests closer to those of future unborn generations. Finally, I discuss the potential for UN human rights mechanisms having a greater role in relation to climate change policy-making (7.7). Also assessed are the advantages and disadvantages of incorporation of human rights discourse in the UN climate negotiations.

7.2 ROLE OF JUSTICE PRINCIPLES IN NEGOTIATIONS

So-called 'realist' approaches to international relations regard reference to justice or fairness in international negotiations as self-serving, with outcomes determined by power relations (Snyder and Diesing 1977).[2] However, burgeoning research across a range of areas of international relations indicates that ethical norms may influence international relations 'not least by shaping the way in which different countries perceive their own interests' (Pickering et al. 2012: 428).[3] Thus fairness can play a significant role in influencing what is feasible.[4] Moreover, empirical evidence suggests that international agreements perceived to be fair amongst parties to the agreement are more likely to be adopted and complied with (Barrett 2003: xiv).

Some have argued that fairness points of agreement will implicitly emerge from negotiations and do not have to be agreed explicitly (Schelling 1960).[5] However, the better view (for example Muller 1999) is that, where there are large differences in perceptions of fairness, bargaining will most likely be successful where negotiators are explicit in identifying focal points: 'when states begin to reconsider and negotiate their own beliefs about what is fair, a mutually acceptable definition will be more likely to emerge' (Roberts and Parks 2007: 220).[6] Successful treaty negotiations tend to move from a starting point of self-serving justice or fairness claims to a 'balanced settlement of conflicting claims' where ideas of fairness and justice influence movement from original positions (Albin 2001: 15, 16).[7]

There is some evidence that the views of fairness (or justice) of the different negotiating blocks in the climate change negotiations reflect countries' economic interests (Lange et al. 2010). For progress in the negotiations to occur, countries will need to move towards a compromise which may embody a number of justice fairness principles. Thus a single notion of fairness need not be accepted by all before agreement can occur. A compromise which entails a 'compromise or negotiated justice settlement' will be sufficient (Roberts and Parks 2007: 150). For example, in the climate change context a compromise could involve moving towards equal per capita emissions over a longer time frame but, in the short term, reflecting a grandfathering approach, combined with the capacity to pay principle (for example Bosetti and Frankel 2011).

7.3 LEGAL FORM

It was argued in 1.9 above that a global treaty regime was essential for addressing the intergenerational justice requirements of climate change: (1) in order to address trade competiveness concerns, (2) to anchor a funding mechanism essential for the development and diffusion of required technology, and (3) to maximise compliance. Further reasons for favouring a treaty-level instrument may be added to this list.

We have seen that from an intergenerational justice point of view, 'effectiveness' is vital (3.4 above) to avoid future generations suffering unnecessary harm. An *effectiveness imperative* (EI) was defined, linked to the notion of 'dangerous anthropogenic emissions' based on Article 2 of the UNFCCC (3.4.1 above). To meet this *effectiveness imperative*, emission reduction targets are required as well as a system for ensuring compliance with such targets. In general terms, agreements binding under international law (treaties) as opposed to non-binding instruments are more likely to be complied with as the reputational costs of non-compliance are higher than compared with breach of a political undertaking (Bodansky 2012b: 3). An effective global climate regime depends on embedding both short- and long-term emissions targets in a global treaty to maximise the likelihood of compliance. From the point of view of future generations, a key advantage of this approach is that commitments binding under international law are likely to be more stable over the long term. Non-treaty-status undertakings may be more easily abandoned following, for example, a change of government.

However, there is a trade-off between ambition (in terms of mitigation) and buy-in. A non-legally binding agreement based on the Copenhagen pledge and review model may be faster to negotiate and attract wider support more quickly (Bodansky 2012b: 2). But the effectiveness of a climate treaty depends not only on participation but also on the stringency of commitments and compliance (Bodansky 2012b: 1).[8] In working to implement intergenerational justice in an effective global climate regime, lessons can be learned from other treaty regimes.

An analysis of effective international treaties, such as those establishing the European Court of Human Rights and the World Trade Organization (WTO) dispute settlement system, suggests that these regimes evolved over a considerable period of time and depended upon the development of trust (Bodansky and Diringer 2010). The dilemma, in terms of intergenerational justice, is that time is of the essence. So the time essential for building trust in a global treaty regime is restricted.

In the environment field, some lessons for negotiating a global climate regime can be learned from the Montreal Protocol on Ozone Depletion (Barrett 2007: 74–102). Comparisons here, however, must be made with care as the Montreal Protocol involved regulation of chemicals with a limited range of uses and the cost of substitute chemicals was easily affordable. By contrast, climate change involves regulating GHG emissions linked to a much broader range of economic activities with much higher costs. Thus the negotiation of a climate regime is much more complex (Barrett 2007: 74–102). How has the Montreal Protocol dealt with compliance issues? The Montreal Protocol has an Implementation Committee, but this body is only entitled to make recommendations rather than decisions (Lesniewska 2010: 478). Non-compliance issues have become more acute as the Montreal Protocol has become more complex, with increasing problems in relation to the 'reporting of data', 'non-fulfilment of control schedules and timetables' albeit linked to a lack of capacity amongst developing countries (Lesniewska 2010: 480). In spite of problems with illegal trade in ozone-depleting substances (UNEP 2007), the overall effectiveness of the Montreal Protocol remains strong, with a 2010 scientific Assessment Panel report finding that globally the ozone layer is projected to recover to its 1980 level around the middle of this century (UNEP 2010: ES5). The Montreal Protocol offers important lessons in terms of the carrots and sticks required to ensure participation (Barrett 2007: 81). 'Carrots' took the form of a funding mechanism to pay the costs of implementation for developing countries. This played a major role in large developing countries such as China and India joining the regime (Lawrence 1990). 'Sticks' took the form of a ban on trade in ozone-depleting substances with non-parties. This also played a significant role in ensuring broad participation in the regime as staying outside the regime involved considerable costs (Barrett 2007: 83).

In the period prior to the Durban COP in 2011, there were conflicting views amongst states as to whether a global climate agreement should be legally binding based on the Kyoto model, with economy-wide emission reduction targets, or rather should be based on the Copenhagen pledge and review model. In brief, the EU, Association of Small Island States (AOSIS) and some others supported a top-down global treaty on the lines of Kyoto. China, India and many other developing countries argued that it was premature to consider a global treaty until at least the IPCC review was completed in 2013 (Rajamani 2012: 504).

This controversy was not completely resolved in the Durban platform (above 4.2.1). The reference to an agreement 'with legal force' could be interpreted as including an agreement which coordinated commitments under national law, following the Copenhagen pledge and review model (Rajamani 2012: 506).

The time frame built into the Durban platform for negotiating a global agreement (2015 with entry into force in 2020) is at odds with the requirements of intergenerational justice as the *effectiveness imperative* can only ever partially be met (3.4 above). While five years' delay in the timescale advocated by scientists may seem small, this delay will entail sharper yearly reductions in emissions being taken later than if stronger action were taken earlier to ensure decarbonisation of the global economy by 2050. It is questionable whether such sharp reductions are feasible (Macintosh 2010). Following release of the IPCC Fifth Assessment Report in late September 2013, pressure should be placed on governments to accelerate the timescales set out in the Durban platform, although it seems unlikely that governments would agree to accelerate this time frame (Rajamani 2012: 516).

7.3.1 The Treaty-making Process

To date, progress in global climate change negotiations has been constrained by dominant economic interests and associated discourses. In addition, the consensus style of the international treaty-negotiation process has tended to make less likely an outcome that protects longer-term interests. According to the consensus-based negotiation process, progress can only proceed as quickly as the most conservative country is willing – a good analogy being a convoy which can only sail as fast as the slowest ship (Bretteville 2001). This allows a small group of countries to block progress in negotiations (Eckersley 2012: 32). The consensus model has also impacted on the presentation of climate change science which underpins the whole policy-making process. The Intergovernmental Panel on Climate Change (IPCC) also adopts a consensus method of operation. As a result of this, the IPCC reports tend to already be out of date by the time that they are publicly released.[9]

Nevertheless, the extraordinary media coverage and increased public concern in relation to the climate change issue have resulted in glimmers of hope in increased pressure on governments to make progress. A good example of this was the US buckling under intense pressure from other countries – and under the spotlight of the international media – to accept the Bali Road Map, which set out a mandate for UN climate negotiations in December 2007.[10]

Eckersley (2012) has proposed that the consensus obstacle can be overcome by restructuring the UNFCCC negotiation process based on 'inclusive minilateralism'. The idea is that key decisions in the negotiation process (such as mitigation targets) should be negotiated in a smaller body representing the 'most capable, most responsible and most vulnerable' (Eckersley 2012: 35), including for example the top 20 GHG emitters responsible for 75 per cent of global emissions, AOSIS, the Africa Group and least developed countries (Eckersley 2012: 32, 35). These groups would be represented in a 'Climate Council', created from the UNFCCC membership, with its decisions conveyed to the COP for approval.

A practical difficulty with this proposal is that it would require a change to the UNFCCC rules of procedure in order to be 'formalised' with a 'clear and fixed remit', which would burn up scarce negotiating time and resources.[11] But a more fundamental difficulty with this approach is its assumption that the failure to reach agreement on an effective global climate treaty is due to the treaty-making process. It is doubtful whether reforming the negotiation process would itself produce agreement. The meagre results to date of international climate diplomacy would seem to have more to do with the nature of the dominant discourses underpinned by powerful economic interests (Chapter 6) than the negotiation process itself.

While the consensus-based treaty-making process is cumbersome, with strong political will, effective global treaties can be negotiated. Examples of effective treaties to date include the WTO dispute settlement system, which has high levels of compliance.[12] In the environment field, examples of effective global treaties include the 1987 Montreal Protocol on Ozone Depletion[13] and the Stockholm Convention on Persistent Organic Pollutants (POPs).[14] These regimes have their problems, particularly in relation to compliance. With strong leadership and the political will, it is possible, albeit difficult, to negotiate an effective climate treaty. An essential ingredient in such negotiations is the inclusion of representative interests in the process, which can be done through, for example, 'Friends of the Chair', where the person chairing the negotiation process brings together key delegations to develop proposals.[15] In the negotiation of the UNFCCC the chairperson of the negotiations, Ripert, used a 'Friends of the Chair' group in the form of an Extended Bureau which included not just major emitters, but countries vulnerable to climate change such as small islands and African countries. This was successful partly owing to trust established between the chair and the participants in the negotiation process, including delegations not included in the Extended Bureau (Borrie and Martin Randin 2005: 77, 83).

7.4 FEASIBILITY CRITERIA

This book has proposed some minimum justice principles which a global treaty must comply with to be considered 'fair' from an intergenerational perspective. Recall that these principles do not stem from a single theory of justice but reflect concepts common to a number of justice theories and widely shared values. Indeed, some of these principles have been strongly endorsed by the international community, for example 'subsistence' in the UN Millennium Development Goals and the core human rights in the UN human rights instruments. These core concepts include core human rights, avoidance of harm, equality and subsistence.

It was argued that *implementation* of these principles required mitigation based on equal per capita shares and sustainable development policies. In the international climate negotiations to date, the equal per capita approach has been supported by the EU, Switzerland and India (Roberts and Parks 2007: 144–6). However, it is opposed by the US, which argues that it would entail prohibitive costs and would be unfair as the US was unaware of the damage occurring through its GHG emissions until 1990 (Roberts and Parks 2007: 144–6).

Does this mean that the principles proposed in this book are unrealistic and incapable of implementation? Not at all. But the design of an effective global treaty must meet political constraints as well as the justice principles. Fully discussing the political constraints for negotiating a global climate agreement is outside the scope of this book. However, Bosetti and Frankel's (2011) model of distribution of mitigation burdens is briefly discussed in order to demonstrate that it is possible for a climate treaty to meet justice requirements within political feasibility constraints.

Bosetti and Frankel's (2011: 3) modelling of appropriate mitigation burdens rests on the notion that 'every country has reason to feel that it is only doing its fair share'. Their model involves, over time, gradually converging to equal per capita emissions. However, US and other countries' fairness concerns related to costs are reflected by no country being asked to accept a path of targets expected to cost more than 1 per cent of income throughout the twenty-first century in current discounted value (Bosetti and Frankel 2011: 7).

The further political constraints proposed by Bosetti and Frankel reflect the positions of key players in the climate negotiations:

1. 'The US will not commit to quantitative targets unless China and other major developing countries commit to quantitative targets at

the same time' (Bosetti and Frankel 2011: 7). This reflects a trade competitiveness concern that without such a commitment major developing countries could exploit their lack of GHG regulation for competitive advantage at the expense of participating industrialised countries who took on emission reductions. It would also address concerns at 'emissions leakage' involving high-emitters moving offshore and thus undermining environmental effectiveness (Bosetti and Frankel 2011: 7).

2. China, India and other major developing countries would not be required to make sacrifices 'different in character from those made by industrialised countries who have gone before them' (Bosetti and Frankel 2011: 7). This would be achieved by, in the short run, developing countries taking on binding targets which coincided with business-as-usual (BAU) emissions. To remove the politics out of assessment of BAU, such assessments would be undertaken by an independent review panel of experts (Bosetti and Frankel 2011: 9). This would address trade competitiveness concerns, as developing countries could not emit GHGs higher than BAU (Bosetti and Frankel 2011: 13).

Pickering et al. (2012: 423) have added a further 'institutional compatibility' constraint: 'A medium-term climate agreement will be feasible only if it maintains a sufficient degree of compatibility with deeply embedded institutional elements of the climate regime.' This constraint reflects the resilience of the UNFCCC and the need to contain elements of its architecture (Pickering et al. 2012: 423). Indeed, a reference to the Article 3 principles of the UNFCCC in a number of the proposals made to date under the Durban platform (below 7.5) shows that it would be wise to respect this 'institutional compatibility constraint' to ensure broad participation in a new agreement.

In order to meet the *effectiveness imperative* posited in Chapter 3, it is essential that at least major developing countries take on emission reduction commitments in the near future, or at least be constrained to BAU as proposed by Bosetti and Frankel (2011). This *effectiveness imperative* suggests that while a new global climate agreement should respect the Annex 1/non-Annex 1 division found in the UNFCCC, this structure needs to be reformed so that the membership of the annexes is based on objective criteria (Pickering et al. 2012). This would address the current anomaly whereby some non-Annex 1 'developing countries' such as Singapore and Qatar have higher per capita GDP than the average developed countries listed in Annex 1 (Winkler et al. 2011: 472).

7.5 EVALUATION OF NATIONAL PROPOSALS FOR A GLOBAL CLIMATE TREATY

Proposals for a global climate regime may be assessed against a range of criteria, including environmental effectiveness, economic efficiency and fairness or justice.[16] As stated at the beginning of this chapter, the criteria for assessment adopted here are argued to be preconditions for ensuring justice for future generations. These criteria are: *core human rights*, *responsibility for harm* and *capacity to pay*, operating in conjunction with core distributional principles of *equality* and *subsistence* under an *effectiveness imperative*. These are augmented by the *political feasibility* criteria of Bosetti and Frankel (presented above) and an *institutional continuity* criteria (Pickering et al. 2012). In order to ensure justice for future generations, proposals must meet the *effectiveness imperative*, *political feasibility* and *institutional continuity* criteria, and substantially the justice principles set out in Chapter 3.

Given that work on negotiating a global climate agreement under the Durban platform has only just begun, there are not, as yet, fully fledged proposals which can be evaluated. However, some countries have set out in broad-brush terms elements to be included in such an agreement. In addition, some proposals made under the Bali mandate are assessed, as these proposals – or elements of them – are likely to be revived in various ways, or at least influence future proposals. The EU internal scheme for allocating emissions is included in the discussion as this scheme may also influence proposals made in the UN negotiations. As already mentioned, the Durban platform remains seriously flawed owing to the slow time frame for negotiations, with serious implications for intergenerational justice.

AOSIS has proposed that a new climate agreement must involve every party making a legal commitment as well as leadership of industrialised countries making emission reductions.[17] The AOSIS proposal reflects well the *institutional continuity* criterion in stating that the principles of the UNFCCC should be reaffirmed and that one should not 'renegotiate the Convention or its annexes'.[18] Developing countries that have not yet done so are urged to make voluntary pledges for emission reductions.[19] AOSIS urges action outside the UNFCCC negotiation process, for example, in reducing fossil fuel subsidies, but argues that this should 'not distract from work in the UNFCCC'.[20] AOSIS, in its presentation to the Bonn workshop on EASD,[21] urged for 'scientifically determined mitigation goals required consistent with well below 1.5°C pathways, with developed countries rapidly reducing emissions 45% by 2020 against

1990 levels and developing countries reducing emissions 15–30% by 2020 against business-as-usual levels but enabled by finance, technology and capacity building'.[22]

The AOSIS submission is strong in terms of the justice principles and *effectiveness imperatives* given the mitigation targets for both developed and developing countries. Where it will need work is on ensuring political feasibility, through finding incentives for all parties to the UNFCCC to take on emission reduction targets, particularly the US, and large developing countries such as India and China.

The EU has proposed[23] that the global climate agreement have a spectrum of commitments 'such that all parties can take on binding ... and ambitious obligations in accordance with their responsibilities and capabilities'.[24] Those with the 'greatest responsibilities and capabilities' are expected to take on 'absolute economy-wide reduction targets'.[25] In addition, the EU proposes that there should be a spectrum of commitments to encourage an as high as possible mitigation effort by all parties, taking into account 'evolving responsibilities, capabilities and circumstances'.[26] The EU has stated that in the long term there is a 'need for gradual convergence of per capita emissions, taking into account national circumstances'.[27]

The EU is committed to a 20 per cent reduction in GHG emissions by 2020 but is willing to raise this to 30 per cent conditional on other countries making comparable commitments.[28] The EU proposal's emphasis on binding commitments for all parties reflects well the *effectiveness imperative* and legal form requirements (7.3 above). However, the current target of 20 per cent by 2020 remains below what is required (1.2 above). The notion of 'evolving responsibilities' reflects well the *responsibility for harm principle* (PII) argued for above (3.5.6). The EU proposal's reference to 'capabilities' reflects well the *capacity to pay* principle (PIII: 3.5.8 above). The EU is yet to flesh out exactly how it would allocate emission reduction quotas within a global agreement. However, the EU's in-principle support for 'convergence towards equal per capita emissions taking into account national circumstances' is consistent with the *modified equal per capita emissions* principle (IP(iv): 3.5.9 above).

While not part of an EU proposal to date, the internal mechanism to allocate emissions within the EU may provide elements which could usefully be incorporated within a global climate agreement, given the wide spectrum of development stages within the EU (Stephenson and Boston 2010). The EU scheme is based on the principles of 'solidarity' and 'capability'. The principle of 'solidarity' means that 'every EU country is obliged to contribute in some way to solving the problem of climate change' (Stephenson and Boston 2010: 4). 'Capability' means that the individual EU member contributions reflect 'relative economic wealth and thus … capacity to help solve the problem' (Stephenson and Boston 2010: 4).[29]

In taking into account each country's individual economic circumstances, the EU's scheme scores well for both protection of *core human rights* (PI: 3.5.1 above) and *political feasibility*, with, for example, poorer EU members being allowed to increase emissions in the short term in order to develop. However, the EU scheme does not reflect the principle of *responsibility for harm* (PII: 3.5.6 above), which may impede its suitability as a model for a global agreement. Nevertheless, Stephenson and Boston (2010: 11) argue that given the broad 'congruence between historical responsibility and capability', the taking into account of capability will at least to some extent reflect historical responsibility. Moreover, in a global agreement, historic responsibility could be taken into account in funding mechanisms which facilitate developing countries' compliance with emission reduction limitations (Pickering et al. 2012). The EU scheme does not directly reflect the implementation principle of *equal per capita emissions* (IP(iv): 3.5.9 above) but is consistent with a 'long-term move towards equal per capita emission rights' (Stephenson and Boston 2010: 11).

The US submission under the Bali mandate reflects in many ways the Copenhagen pledge and review model.[30] The US proposal advocated an implementing agreement under the UNFCCC in which developing country parties would list in Appendix 1 'qualified emission reductions' to be achieved by 2020 and 2050. Developing countries with 'greater responsibility capability' will also have listed 'nationally appropriate mitigation actions' that are quantified.[31] Appendix 1 would be updated based on 'objective criteria of economic development'. Mitigation action is to be subject to 'measurement, reporting, and verification'. The actions of developing country parties are envisaged to be 'supported by financing, technology and capacity building'.[32] To the extent that this proposal reflects specific elements in the Bali mandate, such as the split between developed and developing countries, it may no longer be relevant given the terms of the Durban mandate. This proposal may not meet the requirements of political feasibility by not requiring industrialised countries to pay for mitigation actions of developing countries but merely 'provide support' for such actions. However, in another sense it may be considered politically feasible in the sense that it does not depend upon negotiation of binding mitigation targets contained in a new treaty. But this is also its weakness. From an intergenerational justice point of view, it is unlikely to have the long-term stability of a treaty status instrument and thus fail to meet the *effectiveness imperative*.

The 'Environment Integrity Group' (EIG), comprising Switzerland, Mexico, the Republic of Korea, Lichtenstein and Monaco, have proposed under the Durban mandate that a global climate agreement reflect the objective of Article 2 of the UNFCCC, with 'mitigation at the core', involving 'global participation' and 'common bindingness' and being 'science-based'.[33]

The EIG proposal accords well with the *effectiveness imperative* and 'legal form' requirements required to deliver intergenerational justice. However, it remains to be seen how these countries would propose to distribute mitigation burdens.

Bolivia has made proposals reflecting the notion of 'carbon debt'. The idea is that developed countries owe developing countries a 'carbon debt' by reason of their exceeding their fair share of atmospheric space.[34] This is argued to require 'reductions to zero' by developed countries to enable 'developing countries to have more atmospheric space'. The notion of equal access to atmospheric space and carbon debt reflects well the *responsibility for harm* principle (PII: 3.5.6 above) and is sound from a justice point of view (see above 3.5.4; Pickering and Barry 2012). However, this proposal fails the test of *political feasibility* in placing requirements on developed countries likely to be perceived as unfair by imposing high costs on those countries well beyond, for example, those proposed by Bosetti and Frankel (2011).

Under the Durban platform, Australia has proposed that for an agreement to be effective, '[a]ll major economies should make legally binding mitigation commitments under a legal framework …'.[35] This is to allow countries to take action consistent with 'their respective national circumstances and capacity to act'.[36] Other elements include 'transparency', with 'basic accounting rules that would be common to all parties'[37] and building 'on existing processes and institutions developed under the Convention'.[38] A schedule is proposed to be negotiated to allow non-Annex 1 parties to register a variety of mitigation policies and measures.[39] Included in this proposal is a concept that a party's mitigation commitments should be comparable with other parties at a 'comparable level of development'.[40]

This proposal scores well in terms of *legal form* and *institutional continuity*. However, by not including any element of '*historic responsibility*' it may not meet the justice criteria, which could also in turn impact negatively on its potential *political feasibility*. The proposal does reflect *capacity to pay* by inclusion of references to financial, technology and capacity support being available 'for countries that need it'.[41] This, and inclusion of the 'comparability of effort' principle, may possibly make up for not including historic responsibility, as these two principles may point in the same direction. However, in terms of meeting the *effectiveness imperative*, much will depend on the specific targets involved.

Brazil has made a proposal under the Bali action plan[42] which may be revived in the Durban process or influence other proposals. The Brazilian proposal is strongly based on the notion of *historic responsibility* for Annex 1 countries and climate change,[43] with economy-wide targets advocated for industrialised countries. Non-Annex 1 countries only take on voluntary measures supported through financing and technology transfer by Annex 1 countries.[44] This proposal would reflect well the *responsibility for harm* principle (PII: 3.5.6 above). However, it would not meet the *effectiveness imperative* as major developing country emitters would not be subject to emission reductions. As a consequence of the latter, the US would also be unable to agree, thus failing to meet the test of *political feasibility*. Finally, a lack of reliable data relating to historic GHG emissions would make this proposal unworkable (Höhne et al. 2011).

Indeed, the need to be able to accurately measure GHG emissions points in the direction of counting emissions from, for example, 1990, when the first IPCC assessment report was released (Vanderheiden 2008: 190). Moreover, as mentioned in 1.3 above, the burgeoning growth of developing country emissions over the next decades will mean that pre-1990 emissions will represent a shrinking proportion of global cumulative emissions (Höhne et al. 2011).

Historic responsibility also features prominently in China's proposal under the Bali mandate.[45] Under this proposal, Annex 1 parties are to reduce emissions by at least 40 per cent below 1990 levels by 2020 owing to 'historical responsibility, equity and development stage'.[46] This is 'to allow developing countries development space'. In contrast, non-Annex 1 countries are to make '[n]ationally appropriate mitigation actions' as each country sees fit 'in conformity with the legitimate and prior needs of developing countries for sustained economic growth and eradication of poverty'.[47]

The submission of India under the Bali mandate similarly involves binding emission reduction targets for Annex 1 developed countries with only voluntary mitigation 'actions' for developing countries, the implementation costs of which would be borne by developed countries through a funding mechanism.[48]

Both China and India's proposals reflect well the *responsibility for harm* principle but are not consistent with the *effectiveness imperative*, or *political feasibility* criteria given the failure to include binding commitments on major developing countries.

In summary, then, the proposals of AOSIS and the EU come closest to meeting the *effectiveness imperative*, justice principles, *political feasibility* and *institutional criteria* applied here. Proposals by China, India and Brazil strongly reflect the notions of *historic responsibility* but are weak in terms of the *effectiveness imperative*, owing to a lack of mitigation constraints on major developing country emitters. The US and a number of other countries are likely to continue to refuse to support an agreement that does not include this element, owing to trade competitiveness concerns. As mentioned above, however, the failure of most industrialised countries to make significant reductions in GHG emissions has undermined the basis for developing countries to take on mitigation commitments. Those most seriously impacted by this impasse will be the future poor.

7.5.1 Essential Content of Treaty Commitments

Section 7.3 concluded that only a treaty status instrument is capable of delivering intergenerational justice in relation to climate change. I now turn to address what substantive commitments such a treaty would need to include to meet the justice principles proposed in Chapter 3 within an *effectiveness imperative*. The contours of such a treaty are set out in general terms as my purpose is to demonstrate that such a treaty is feasible. The detail of specific elements, such as emissions reduction targets, would of course be the product of negotiations. The modified equal per capita emission approach advocated above (Chapter 3) should be the basis upon which short- and long-term national emission reduction targets are embedded in a global climate treaty. These emission reduction targets should be quantified within an overall budget requiring zero emissions by 2050. As Bosetti and Frankel (2011) demonstrate, it is feasible to move towards an equal per capita emissions approach gradually, with political feasibility taken into account by each party only having to take a maximum level of emission reductions per year corresponding to a cost cap. We have also seen that the internal scheme operating within the EU for the distribution of GHG emissions involves moving gradually towards an equal per capita emissions approach.

In terms of concrete emission reduction targets, the *effectiveness imperative* dictates that a global treaty include binding emission reduction targets for industrialised countries dictated by climate science, thus set at the level of a reduction of 10–40 per cent by 2020 (against a baseline of 1990) and a 100 per cent reduction by 2050 (1.2 above). The principle of *proportional responsibility for harm* dictates that larger developing countries take on short-term binding constraints of not increasing their emissions beyond BAU, but with these countries gradually adopting significant emission reductions over time. In Chapter 4 we saw that the current climate change regime fails to include emission reduction targets with adequate stringency and coverage given the limited scale of the Kyoto Protocol commitments. Moreover, the time frame for negotiating a new global instrument under the Durban platform remains inadequate.

Intergenerational justice in the climate change context requires implementation of a strong version of sustainable development whereby elements of the global ecosystem which are not substitutable – including the ongoing stability of the global climate – are not traded off against other elements (3.5.2). We further saw that a *structural reform implementation principle* was justified which placed an onus on policy-makers not to delay structural economic reforms necessary to ensure the protection of the long-term interests of future generations while minimising harm to the core rights of current generations (IP(ii): 3.5.3 above).

Should sustainable development and structural reform principles be explicitly incorporated in a global climate treaty and, if so, how? We have seen that the UNFCCC already contains a 'right to promote sustainable development', while this foundation reflected an awkward compromise between developed and developing countries (4.2.8). It would be preferable if a global climate treaty included a duty of strong sustainable development as well as the *structural reform implementation principle*. Both principles – if taken seriously – provide the basis for more specific elements in a global treaty, including in relation to emission reduction targets, technology transfer and compliance.

However, to date, governments have been resistant to the inclusion of treaty commitments which would dictate to national policy-makers how GHG emissions are to be reduced. Thus governments rejected in the negotiations of the Kyoto Protocol inclusion of specific 'policies and measures' to be required to be adopted by governments to reduce GHG emissions (Sands and Peel 2012: 287). This, however, is not an obstacle to an effective global climate treaty as governments can be left free to meet targets in the manner they think fit. Similarly, strong sustainability and structural reform principles do not need to be necessarily embedded in a climate treaty. But a shared understanding on these principles is essential for progress.[49] As discussed in Chapter 4, an effective compliance mechanism is a vital element for an effective global climate treaty. It is beyond the scope of this book to discuss how this element, and other elements such as an effectiveness technology transfer mechanism, should be designed. Such a discussion would need to include the Clean Development Mechanism (CDM) currently operating under the Kyoto

Protocol, which allows industrialised countries to reduce emissions through projects in developing countries. An extension of the CDM in a global instrument offers potential in furthering sustainability goals if properly designed (see Sands and Peel 2012: 288–91).

A global climate treaty of the nature described above could be negotiated under the existing Durban platform, but only if this mandate was amended to bring forward the time frame involved with the new treaty to be finalised by the end of 2014 and entered into force by the end of 2015. This new time frame would accord with the urgency of action required by climate science. As explained above (Chapter 4), there is some ambiguity as to whether the current Durban mandate includes the principles of Article 4 of the UNFCCC, including 'intergenerational equity' and CBDR. For many developing countries – particularly the BASIC group – acceptance of these principles is a prerequisite for ongoing negotiations (Chapter 4). A deal needs to be struck whereby industrialised countries urgently make sharp emission reductions and fund the costs of major developing country emission reductions. In return, major developing countries would accept a notion of proportionate responsibility entailing emission reduction targets. The deal sketched above would not depend on *explicit* recognition of the justice principles set out in Chapter 3 above. But a compromise reflecting a combination of these principles is required. Responsibility for historic emissions remains a sticking point in reaching such a compromise. As proposed by Pickering et al. (2012), developed countries' historic emissions could be reflected in a global treaty's funding mechanism but not in the assignment of emission reduction targets, as the measuring of historic emissions is too unreliable (Chapter 1 above).

7.6 IMPLEMENTING PROCEDURAL JUSTICE FOR FUTURE GENERATIONS

7.6.1 Reforming the International Negotiation Process

We have seen that procedural justice requires that those who are impacted by a particular decision should have input into the making of that decision (1.5 above). The rationale for this is that with such participation there is a greater chance of outcomes which meet the requirements of substantive justice. Indeed, procedural justice mechanisms only have value to the extent that they deliver substantive justice. A thorough analysis of procedural mechanisms which could ensure greater representation of future generations' interests in climate treaty-making is important but outside the scope of this book. Of course, it is impossible for future unborn generations to have direct input into the current UN climate change negotiation process. However, young people alive today share to a significant extent the interests of future unborn persons in relation to climate change impacts. Moreover, young people alive today, within their own lifetimes, will be impacted by climate change long after those in positions of power and authority have gone. Thus there are compelling reasons for giving young people a strong voice in climate policy-making.

However, young people currently have minimal opportunities to influence climate change policy-making. We have seen that youth climate change NGOs currently participate in the UN climate change negotiations, but only as observers with no right to participate directly in negotiations (Chapter 4). How could this be changed to better reflect procedural justice imperatives?

One option is to create new mechanisms that give younger people a greater say in the negotiation process. This could be done at both the national and international level. Options worth consideration include incorporation of youth representations on government delegations, the creation of a United Nations Climate Change Youth Ambassador Programme, and the creation of national Ombudspersons for Future Generations.

There is nothing preventing national delegations including youth representations on delegations. Indeed, some governments already undertake this practice. Another possibility is expansion of the current United Nations Youth Ambassadors programme.[50] Only about 20 countries (mostly continental European) have the position of Youth Representatives to the UN and there is potential to further develop this programme. This could well be done by creating a specific 'United Nations Climate Ambassadorship Programme' aimed at providing increased opportunities for youth around the world to express their views in the ongoing UN negotiation process. Ideally, such climate youth ambassadors should be given dedicated space on the UNFCCC meeting agenda. While the current Durban mandated agenda is daunting in its complexity, as a minimum, national delegations could be encouraged to give regular briefings to such youth ambassadors. The programme could be based on voluntary funding with a twinning programme whereby an industrialised country had a partnership with a developing country party, with the industrialised country paying for the related travel costs to facilitate attendance at meetings.

A further possibility would be the appointment of an Ombudsman for Future Generations. At the international level, a Commissioner for Future

Generations was considered in the negotiations leading up to the Rio+20 Conference held in Brazil in 2012. However, as discussed in Chapter 1, this proposal fell out of the final text, which only acknowledges that opportunities to participate in decision-making are 'fundamental to sustainable development'.[51]

Indeed, this particular proposal is the most recent of a string of similar proposals over recent decades. These have included a proposal for a 'guardian for future generations', discussed during the 1992 UN Rio Conference on Environment and Development (Redgwell 1999: 89). Such proposals seek to build on similar mechanisms at the national level. Israel, for example, created a Commissioner for Future Generations which operated from 2001 until 2007.[52] Reform of the United Nations Trusteeship Committee has been proposed, entailing the idea that following completion of the decolonisation process the Committee could exercise trusteeship for the 'integrity of the global environment and common areas such as the oceans, atmosphere and outer space'.[53] This latter proposal could include climate change, but would entail amendment of the UN Charter (Redgwell 1999: 89).[54] To date there has not been the political will to undertake such a step. The 1992 Rio Conference did lead to the creation of the Commission on Sustainable Development, but this body has a role in monitoring environmental programmes rather than acting on behalf of future generations.[55]

However, whether the procedural reforms set out above will actually deliver *substantive* intergenerational justice remains uncertain and there is a need for further research in this area. Given that by definition sustainable development incorporates the interests of future generations, careful thought needs to be given as to whether mechanisms at the international or national level purporting to represent future generations unnecessarily duplicate the mandates of existing institutions which have a sustainable development focus.[56] Moreover, there is a risk that if such mechanisms are created, governments take the view that intergenerational justice issues are therefore fully addressed. It is clear, however, from the analysis of dominant discourses (Chapter 6), that a shift is required in these discourses as a precondition for the implementation of intergenerational justice principles. Ethics and justice discourses need to become more prominent in climate policy-making. The discourse of ecological modernisation which contains within it a notion of intergenerational justice needs to be fully implemented (7.1 above). In addition, strategies are required to combat dominant economic interests that underpin those discourses.

7.7 HUMAN RIGHTS OPTIONS

Human rights, by attributing equal value to persons regardless of when they happened to be born, can counter powerful economic discourses, some strands of which discount the interests of future generations (1.7). We have also seen that human rights litigation faces acute obstacles as a vehicle for addressing intergenerational justice aspects of climate change (Chapter 5). This section considers whether justice for future generations in relation to climate change would be furthered by: (1) climate change mitigation issues being taken up more extensively within the UN human rights system, and (2) human rights discourse being incorporated within the UNFCCC negotiations. The latter issue is considered by discussing whether human rights provides a valuable benchmark for decisions on appropriate mitigation targets.

In relation to the first issue, it is important to note that within the UN human rights system, conflict remains between industrialised and developing countries on the importance of civil and political rights, compared with that of economic and social rights, with developing countries, particularly China, emphasising the latter (see 2.1 above). This conflict has lessened in recent years with the Obama administration now giving qualified support to economic and social rights (Alston and Goodman 2013: 293).

There has also been ongoing North–South conflict in relation to the right to development, with the US in particular opposing such a right (Shaw 2008: 302). The Obama administration has softened its opposition to the right to development but still opposes its adoption as a legally binding principle.[57] Moves to link human rights and climate change governance must be undertaken carefully to avoid worsening these conflicts.

There is a real danger that a push for the incorporation of human rights discourse into the UN climate change negotiation process could further exacerbate the existing North–South conflicts within this process and burn up precious negotiating time, with little to show in return (Pederson 2011: 250).[58] While most countries in the world now give formal support for the universality of human rights, China, for example, still perceives UN human rights machinery as a vehicle for Western countries unfairly ‘pointing the finger’ at China (see 2.1). Moreover, there is considerable ‘ideological baggage’ relating to the relationship between economic social and cultural rights and civil and political rights which has manifested itself in North–South conflict in UN human rights forums over recent decades (Alston and Goodman 2013: 278).[59] This suggests that importing human rights discourse into the UN climate negotiations necessarily would bring with it the considerable 'ideological baggage' of this North–South conflict. This would seem to conflict with my argument in Chapter 2 that the core human rights to life, health and subsistence provide a sound basis for a moral obligation extending to future generations on the basis that these human rights reflect widely shared values. This apparent contradiction is explained in that, while these core human rights do reflect widely shared values, *any* introduction of human rights discourse in the UN climate change negotiations would necessarily risk bringing with it tensions related to the value of economic, social and cultural rights vis-à-vis other human rights and a broader raft of human rights issues including freedom of religion, freedom of expression and rights to democratic participation which continue to be perceived by at least some non-Western countries as being inextricably linked to Western liberalism. The upshot of this is that any use of human rights discourse in the climate change negotiations could risk raising hackles rather than improving North–South cooperation.

On the other hand, there does seem value in the UN Human Rights Council continuing to point out the link between the failure to address climate change and the infringement of human rights. The incorporation of climate change impacts into existing human rights mandates, such as that of the Special Rapporteur on the Right to Food, is welcome in drawing attention to the breadth of human rights impacts which climate change will increasingly have on vulnerable groups.

Some have suggested going beyond this to create a new mechanism, for example a Special Rapporteur on Human Rights and Climate Change. However, this approach would be potentially problematic if it involved entering into policy areas better addressed by the UNFCCC.

To illustrate this point, consider the key issue in the UN climate change negotiations: deciding on specific greenhouse gas emission reduction targets for countries or groups of countries within a new global climate agreement. The setting of emission reduction targets necessarily involves policy decisions to be made about the emission reductions which entail distributional issues, both in terms of climate change mitigation burdens and climate change impacts between countries and between current and future generations. As we have seen, issues of justice are integral to the climate change negotiations. Decisions by governments on reduction targets will require input from: scientists; economists, in terms of the most efficient design of targets; industry, in terms of technology options; as well as NGOs and others. A Special Rapporteur on Human Rights and Climate Change could broadly highlight the need to take strong mitigation action to ensure the protection of the rights of current and future generations. But beyond this, it is difficult to see what contribution would be made by creating such a position. Broad statements about the need to negotiate a strong regime to protect the human rights of current and future generations would arguably add nothing to the existing UN Human Rights Council resolutions discussed above. For a Special Rapporteur to go beyond this and address substantive issues in the negotiations by making specific proposals on the elements to be included in an international regime would surely just complicate already very complex and difficult negotiations.

A Special Rapporteur on Human Rights and Climate Change could potentially add value by acting as a mediator or 'friend of the chair' in facilitating the UNFCCC negotiations. However, this would depend on the individual involved having the necessary characteristics and trust of delegations to perform this role. While there is no shortage of high-profile persons capable of performing such a role, the slim possibility of a Special Rapporteur carrying off such a role would not justify the considerable expense in funding such a position and associated resources in negotiation of the mandate of such a Special Rapporteur (cf. Knox 2009: 496–7).[60]

The Human Rights Council in its March 2012 session created the position of an Independent Expert on Human Rights and the Environment[61] and appointed John Knox – Professor of International Law at the School of Law of Wake Forest University – to this position.[62] The creation of this position had a precedent with a Special Rapporteur on Human Rights and the Environment – Fatma Zohra Ksentini, reporting to the Human Rights Commission in 1994,[63] following an earlier report on toxic waste and the environment.[64] It remains unclear what impact these reports have had in practical terms. The mandate of the newly created position includes promoting and strengthening environmental policy-making including preparing 'a compendium of best practices' (para 2(b)) but there is no specific mandate to become directly involved in the climate change negotiations. It will be interesting to see how this mandate is implemented and whether it weakens or strengthens the push to create a specific Special Rapporteur on Human Rights and Climate Change.

While human rights provides a solid basis for ethical obligations towards future generations in the climate change context, human rights cannot in itself resolve key policy conflicts, such as those between the rights of current and future generations. Distributional justice is required to address this issue and we saw that the capabilities approach of Sen and Nussbaum has great potential in this regard, in spite of some limitations (2.4). As a benchmark for fleshing out climate change mitigation targets, human rights is of limited value in that its prescriptions are overly vague, adding little to existing UNFCCC obligations. However, human rights entails the idea that persons are equal regardless of when they were born, which can help counter powerful economic discourses, some strands of which discount future damage and implicitly the interests of future generations.

7.8 CONCLUSION

The evaluation of national proposals under the Durban platform shows that there remain sharp divisions between various groups of countries and in particular between larger developing countries and the US on how to fairly distribute mitigation burdens. This ongoing impasse on what *intra*generational justice means has potentially dire consequences for future generations. Negotiation literature suggests that an explicit agreement on justice principles is not necessarily required. A compromise based on competing justice principles would be sufficient. Such a compromise depends on developing trust within the UNFCCC regime by industrialised countries taking actions to sharply reduce GHG emissions. In other forums, such as the WTO, industrialised countries need to take action to address developing countries' requirements for development.

In spite of the enormous challenge of reaching agreement on a global climate regime, this chapter has argued that international law, in the form of a treaty with a strong compliance mechanism, remains essential to address trade competitiveness concerns. Moreover, only treaty-level commitments can provide the long-term stability necessary to address the interests of future generations. The model developed by Bosetti and Frankel (2011) and the scheme operating within the EU to distribute mitigation burdens both suggest that it is feasible – albeit very difficult – to implement the justice principles proposed in this book in a global climate agreement.

More specifically, this chapter argued that a global climate treaty requires binding short- and long-term emission reduction targets for industrialised countries and developing countries allocated within an emissions budget which delivers zero emissions by 2050. These targets should gradually reflect the equal per capita principle advocated above (3.5.9). Historic emissions by industrialised countries would be reflected in the funding mechanism but not emission reduction targets owing to difficulties in accurately measuring historic emissions.

The Durban mandate which establishes the current mandate for a global climate instrument needs to be urgently amended to include a tighter time frame, with a new global treaty to enter into force by 2015. A deal is required whereby larger developing countries accept binding emission reduction targets in return for developed countries sharply reducing their emissions and providing funding for developing countries. This would entail a new understanding of common but differentiated responsibilities based on proportionate responsibility as part of a justice compromise which need not be explicitly recognised.

While human rights provides strong ethical grounds for action to protect future generations, giving the current UN human rights mechanisms a stronger voice in relation to climate change policy-making needs to be handled with care. Moreover, an injection of human rights discourse into the UNFCCC negotiation process could be counter-productive and exacerbate existing North–South tensions.

In relation to procedural justice, this chapter has argued that there is an urgent need to expand the possibilities for the interests of future generations being taken into account in the climate treaty-making process. One modest proposal to expand such possibilities is the creation of a specific UN youth climate ambassadors programme. While outside the scope of this book, there is a need for further research on both international and national mechanisms which incorporate the representation of young people at the national and international level in climate change treaty-making in order to ensure that such policy-making represents better the interests of youth and future unborn generations. However, it is vital that such proposals be developed carefully and be genuine, not tokenistic.[65] Such procedures, even if implemented, would not necessarily address the need for a shift in the dominant discourses (Chapter 6) essential for ensuring substantive justice for future generations. As a minimum, ecological modernisation, involving the incorporation of environmental damage in the costs of goods and services, needs to be fully implemented (7.1).

NOTES

1. Based on economics, ethics or science. See surveys in: Aldy et al. (2003), Victor (2004) and Stern (2007).
2. See also references in Albin (2001: 5–7).
3. Pickering et al. refer to Reus-Smit and Snidal (2008). (Reus-Smit C and Snidal D, *The Oxford Handbook of International Relations* (Oxford University Press 2008)). See also literature referred to in Albin (2001: 7).
4. Pickering, Vanderheiden and Miller (2012: 428).
5. Referred to in Roberts and Parks (2007: 220). (Schelling TC, *The Strategy of Conflict* (Harvard University Press 1960)).

6. Roberts and Parks note that their research supports Muller's (1999) conclusions on this point (Roberts and Parks 2007: 220).
7. Albin (2001: chapter 3) demonstrates this process in a case study involving a regional treaty to address acid rain.
8. See also Barnett (2003).
9. See for example Paul Eccleston, 'Climate change accelerating far beyond IPCC forecast, WWF says', 20 October 2008, Telegraph.co.uk, www.telegraph.co.uk (accessed 13 January 2009).
10. United Nations Climate Change Conference in Bali, 3–14 December 2007, Bali, Indonesia, *The Earth Negotiations Bulletin*, IISD Reporting Services, www.iisd.ca – International Institute for Sustainable Development, accessed 13 January 2009.
11. In spite of years of negotiations, the UNFCCC parties have to date been unable to agree on Rules of Procedure, with views divided on the formula for majority voting where a consensus cannot be reached (Eckersley 2012: 34).
12. Valerie Hughes, 'The Institutional Dimension', chapter 11 in Bethlehem et al. (eds) (2009: 272–3).
13. In force 1 January 1989, 26 ILM 154 (1987).
14. Convention on Persistent Organic Pollutants (Stockholm) 2001, in force 17 May 2004, 40 ILM 532 (2001). This Convention appears to be operating successfully in identifying and regulating new POPs of international concern. However, it remains to be seen how this regime responds to 'moves to regulate POPs whose toxicity is not uniformly accepted, and for which the socio-economic consequences of bans would be more acute for many countries' (Sands and Peel 2012: 526).
15. While the use of this mechanism can lead to the exclusion of countries who then block a resulting consensus, as happened in Copenhagen, this does not have to be the case.
16. These criteria are derived from Gupta et al. (2007: 747), who point out that climate change policy has been evaluated against the criteria of 'environmental effectiveness, cost-effectiveness, distributional effects (including equity) and institutional feasibility'. These criteria subsume more specific ones, including 'effects on competitiveness and administrative feasibility'.
17. Submission by AOSIS to the Ad Hoc Working Group on the Durban Working Platform for Enhanced Action. http://unfccc.int/bodies/awg/items/6656.php (accessed 1 February 2013) 4.
18. Ibid., 4.
19. Ibid., 6.
20. Ibid., 7.
21. Presentation of AOSIS to the workshop on EASD, Bonn, 16 May 2012, http://unfccc.int/files/bodies/awg-lca/application/pdf/20120516 (accessed 9 January 2013).
22. Ibid., slide 7.
23. Submission by the European Union to the Ad Hoc Working Group on the Durban Working Platform for Enhanced Action. http://unfccc.int/bodies/awg/items/6656.php (accessed 1 February 2013).
24. Ibid., 3.
25. Ibid., 3.
26. Ibid., 3.
27. Presentation of the European Union at the Workshop on EASD, 16 May 2012, Bonn http://unfccc.int/files/bodies/awg-lca/application/pdf/20120516_eu_0924.pdf (accessed 4 November 2012).
28. FCCC/KP/CMP/2012/L.9 (accessed 8 December 2012) 8.
29. The EU scheme involves apportioning emissions using different principles for three sectors: 'power generation, energy-intensive industry' and 'emissions from all other activities including domestic, commercial, transport and agriculture' (Stephenson and

Boston 2010: 4). The suitability of a sectoral approach for a global climate regime is beyond the scope of this book.

30. 'Ideas and proposals on the elements contained in paragraph 1 of the Bali action plan' submissions from parties Part I (19 May 2009) FCCC/AWGLCA/2009/MISC. 4 (PART 2) 106.
31. Ibid., 107.
32. Ibid., 108.
33. Environmental Integrity Group (EIG) proposal to the Ad Hoc Working Group on the Durban Platform for Enhanced Action November 2012, http://unfccc.int/bodies//awg/items/6656.php (accessed 1 February 2013) 1.
34. Presentation of Bolivia at the workshop on EASD, 16 May 2012, Bonn, www.unfccc.int/meetings/bonn_may_2012/workshop/6658.php(20120515_bolivia_2340.pdf) (accessed 15 January 2013), 5.
35. Australia's proposal to the Environmental Integrity Group (EIG) proposal to the Ad Hoc Working Group on the Durban Platform for Enhanced Action November 2012, http://unfccc.int/bodies//awg/items/6656.php (accessed 1 February 2013) 3.
36. Ibid., 3.
37. Ibid., 3.
38. Ibid., 4.
39. 'Ideas and proposals on the elements contained in paragraph 1 of the Bali action plan' submissions from parties Part I (19 May 2009) FCCC/AWGLCA/2009/MISC.4 (PART I).
40. Ibid., 22.
41. Australia's proposal to the Ad Hoc Working Group on the Durban Platform for Enhanced Action November 2012, http://unfccc.int/bodies/awg/items/6656.php (accessed 1 February 2013) 3.
42. 'Ideas and proposals on the elements contained in paragraph 1 of the Bali action plan' submissions from parties Part I (19 May 2009) FCCC/AWGLCA/2009/MISC.4 (Part I) 54–5.
43. Ibid., 54.
44. Ibid., 55.
45. 'Ideas and proposals on the elements contained in paragraph 1 of the Bali action plan' submissions from parties Part I FCCC/AWGLA/2009/MISC. 4 (Part I) 63–4.
46. Ibid., 63.
47. Ibid., 64.
48. 'Ideas and proposals on the elements contained in paragraph 1 of the Bali action plan' submissions from parties Part I FCCC/AWGLA/2009/MISC. 4 (Part I) 107.
49. See Brunnee and Toope (2010: chapter 4) for the importance of developing shared understandings as a basis for building an effective global climate change treaty.
50. http://www.unyouth.org.au/representation/youthrep/. The experience of Ben Groom, Australia's Youth Ambassador to the UN in 2007, is interesting in this respect. Groom visited more than 100 schools and 34 cities talking to young people before addressing the General Assembly in October 2007 (personal communication with Ben Groom, 4 July 2008).

    Groom took up the issue of climate change in his address, arguing for the need for strengthening education and training. When I asked how much value there was in the inclusion of young people in such positions, Ben said that young people could influence older generations through presence around the family dinner table to push for a switch, for example, to renewable sourced energy. As he said in his speech to the UNGA: 'Young people are the most effective agents of change.' (Ben Groom, *Speech to the United Nations General Assembly*, 9 October 2007, http://bengroom.youthrep.org.au-Ben Groom-Australia Youth Representative to the United Nations, accessed 13 January 2009, at 1.)

51. 'Report of the United Nations conference on Sustainable Development', Rio de Janerio, Brazil, 20–22 June 2012 (A/CONF.216/16) (2012) para 13.
52. Knesset Law (Amendment No. 14), 5761-2001 unofficial translation reprinted in *The Knesset, Commissioner for Future Generations* 13–20 (2004), discussed in: Science and Environmental Health Network and The International Human Rights Clinic at Harvard Law School, *Models for Protecting the Environment for Future Generations*, October 2008, www.law.harvard.edu/programs/hrp/documents/Models-four-Protecting-the-environment-for-Future-Generations_1r).pdf, (accessed 16 January 2009) at 18.
53. Report of the United Nations Secretary-General, 'Renewing the United Nations: A Programme for Reform', UN Doc. A/51/950, para 85 (14 July 1997), referred to in Redgwell (1999: 88–9).
54. For a good overview of trustee-type mechanisms in international environmental law, see Sand (2004: 12).
55. Ibid., 88.
56. See Victoria's Commissioner for Environmental Sustainability, www.ces.vic.gov.au/ces/wcmn.nsf.
57. See US Explanation of vote, United Nations General Assembly, Third Committee, Resolution on the Right to Development (22 November 2010) www.state.gov/documents/organization/179241.pdf (accessed 2 August 2012).
58. Pederson (2011: 250) also makes the argument that human rights may reinforce 'unhelpful polarisations' and in its use of the language of 'violators' and 'victims' is unhelpful given that we are all, in the climate change context, 'to some extent, violators and victims'.
59. Alston and Goodman (2013: 278) observe that the debate on the relationship between civil and political rights on the one hand and economic, social and cultural rights on the other has since the 1970s taken on 'important North–South dimensions' including 'claims that developing countries should not be held to the same standards in some respects and that respect for rights by poorer states must be linked to international aid, and trade and other concessions. As a result, the debate carries a lot of ideological baggage.'
60. See letter by NGOs calling for the Human Rights Council to establish a Special Procedure – such as a Special Rapporteur on Human Rights and Climate Change, 28 February 2011, http://iefworld.org/node/307 (accessed 3 August 2012).
61. United Nations General Assembly UN Doc. A/HRC/19/L.8/Rev.1 (20 March 2012).
62. 'Appointment of Special Procedure mandate-holder, 34th Meeting, 6 July 2012' http://www.unmultimedia.org/tv/webcast/2012/07/appointment-of-special-procedure-mandate-holder-34th-meeting.html (accessed 3 August 2012).
63. Review of further developments in fields with which the sub-commission has been concerned with human rights and the environment Final report prepared by Mrs Fatma Zohra Ksentini, Special Rapporteur E/CN.4/Sub.2/1994/9 (6 July 1994) http://www.unhchr.ch/Huridocda/Huridoca.nsf/0/eeab2b6937bccaa18025675c005779c3?Opendocument (accessed 26 September 2012).
64. There have been ongoing Special Rapporteurs on toxic wastes and human rights since the mid-1990s. See Office of the High Commissioner for Human Rights, United Nations Human Rights, 'Special Rapporteur on the implications for human rights of the environmentally sound management and disposal of hazardous substances and wastes' (accessed 4 October 2012) http://www.ohchr.org/EN/Issues/Environment/ToxicWastes/Pages/SRToxicWastesIndex.aspx.
65. Groom (Australia's Youth Ambassador to the UN in 2007 – see above n49) said that he had full discretion in terms of the content of his speech to the UN General Assembly. However, the website for the Australian Youth Ambassador Program states in small print at the bottom 'DFAT [Department of Foreign Affairs and Trade] reserves the right to withdraw the Youth Representative position at any time.'

Groom's speech is at Ben Groom, Speech to the United Nations General Assembly, 9 October 2007, http://bengroom.youthrep.org.au-Ben Groom-Australia Youth Representative to the United Nations, (accessed 13 January 2009), at 1.

8. Conclusion

This book has argued that an effective climate change regime which ensures justice for future generations is possible if built on acceptable combinations of justice-based principles. To attain such an agreement it is essential to identify shared understandings of justice and ethics. Central in the justice-based framework developed in this book have been notions of core human rights, equality and subsistence. We have seen in the analysis of climate change discourses that normative arguments are integral to countries' negotiating positions and these in turn are underpinned by powerful economic interests.

Both of the above factors suggest that analysis of justice and ethics in relation to climate change is not just a theoretical philosophical exercise but a practical and pragmatic task. A key element of the approach of this book has been to seek to understand international climate law in its political and economic context to ensure reform proposals are well grounded. This approach is dictated by the inherently multidisciplinary nature of the problem of climate change.

In spite of the obvious multidisciplinary nature of the climate change problem, much of the literature on climate change has been undertaken in silos, whether the silos be those of the international lawyer, economist, philosopher or international relations scholar. The risk of such narrow approaches is that they result in either hopelessly utopian proposals, irrelevant to the real world, or alternatively that they are based – explicitly or implicitly – on ethical assumptions which may be difficult to justify.

Partly owing to the domination of the positivist tradition in international law, the use of justice-based frameworks to critique international law has seldom been undertaken in the recent period.[1] Unsurprisingly, new research questions have become evident in applying such an approach to the issue of intergenerational justice and climate change mitigation. These include the crucial issue of how to combine justice-based principles in a politically feasible climate treaty. This book seeks to demonstrate that it is difficult but possible to attain such an agreement. But much further research work is required to address this fundamental issue.

To date, the international climate regime has had little success. Some may therefore regard this book as overly optimistic in its emphasis on the need for and feasibility of a climate treaty, binding under international law. However, a pessimistic, defeatist view is likely to become a self-fulfilling prophecy. Optimism tempered with realism is the most justifiable approach. There is much to be done and no time to lose.

NOTE

1. Notable exceptions are Franck (1995) and Buchanan (2004).

Bibliography

ABC News, 'Govt encouraged to welcome "climate change refugees"' (14 July 2008), http://www.abc.net.au/news/stories/2008/07/14/2302737.htm (accessed 19 December 2008).

Agarwal A and Narain S, *Global Warming in an Unequal World: A Case of Environmental Colonialism* (Centre for Science and Environment 1991).

Agius E and Busuttil S (eds), *Future Generations and International Law* (Earthscan 1998).

Albin C, *Justice and Fairness in International Negotiation* (Cambridge University Press 2001).

Aldy J, Barrett S and Stavins R, '13+1: A Comparison of Global Climate Architectures' (2003) 3(4) *Climate Policy* 373.

Alford R and Tierney J, 'Moral Reasoning in International Law', in DE Childress (ed.), *The Role of Ethics in International Law* (Cambridge University Press 2012).

Allen MR (2009a), et al. 'Warming Caused by Cumulative Carbon Emissions Towards the Trillionth Tonne' (30 April 2009) 458 *Nature* 1163.

Allen M (2009b), et al. 'The Exit Strategy' (May 2009) 5 *Nature Reports Climate Change* 357.

Allison I, et al., *The Copenhagen Diagnosis: Updating the World on the Latest Climate Science* (The University of New South Wales Climate Change Research Centre 2009).

Alston P and Goodman R, *International Human Rights* (Oxford University Press 2013).

Anderegg W, 'Diagnosis Earth: The Climate Change Debate' (2010) *Thought & Action* 23.

Anderegg W and Harold J, *Climate Science and the Dynamics of Expert Consensus* (Schneider Lab, Center for Conservation Biology, Stanford University 2009).

Armstrong D, Farrell T and Lambert H, *International Law and International Relations* (2nd edn, Cambridge University Press 2012).

Atapattu S, 'Climate Change, Differentiated Responsibilities and State Responsibility: Devising Novel Legal Strategies for Damage Caused by Climate Change', in BJ Richardson, et al. (eds) *Climate Law in Developing Countries* (Edward Elgar 2009).

Australian Academy of Science, *The Science of Climate Change – Questions and Answers* (Australian Academy of Science 2010).

Baer P, 'Equity, Greenhouse Gas Emissions, and Global Common Resources', in S Schneider, A Rosencranz and J Neiles (eds), *Climate Change Policy: A Survey* (Island Press 2002).

Baer P, 'Adaptation to Climate Change: Who Pays Whom?', in W Adger, et al. (eds), *Fairness in Adaption to Climate Change* (MIT Press 2006), reprinted in S Gardiner, et al. (eds), *Climate Ethics: Essential Readings* (Oxford University Press 2010).

Baer P, with Athanasiou T, Kartha S and Kemp-Benedict E, 'Greenhouse Development Rights: A Framework for Climate Protection That Is "More Fair" Than Equal Per Capita Emissions Rights', in S Gardiner, et al. (eds), *Climate Ethics: Essential Readings* (Oxford University Press 2010).

Barnett J, 'Security and Climate Change' (2003) 13 *Global Environmental Change* 7–17.

Barrett S, *Environment and Statecraft: The Strategy of Environmental Treaty-Making* (Oxford University Press 2003).

Barrett S, *Why Cooperate? The Incentive to Supply Global Public Goods* (Oxford University Press 2007).

Barry BM, 'Justice Between Generations', in P Hacker and J Raz (eds), *Law, Morality, and Society* (Clarendon Press 1977).

Barry B, *Theories of Justice* (University of California Press 1989).

Barry B, 'Sustainability and Intergenerational Justice', in A Dobson (ed.), *Fairness and Futurity: Essays on Environmental Sustainability and Social Justice* (Oxford University Press 1999).

Bartlett S, *Environment and Statecraft: The Strategy of Environmental Treaty-Making* (Oxford University Press 2003).

Bartlett S, *Why Cooperate? The Incentive to Supply Global Public Goods* (Oxford University Press 2007).

Bastmeijer K, 'Intergenerational Equity and the Antarctic Treaty System: Continued Efforts to Prevent "Mastery"' (2011) 3 *Yearbook of Polar Law* 635.

Bates G, *Environmental Law in Australia* (6th edn, LexisNexis 2006).

Becker LC, *Reciprocity* (The University of Chicago Press 1986).

Beckerman W, 'Sustainable Development and our Obligations to Future Generations', in A Dobson (ed.), *Fairness and Futurity: Essays on Environmental Sustainability and Social Justice* (Oxford University Press 1999).

Beckman L, 'Do Global Climate Change and the Interest of Future Generations Have Implications for Democracy?' (2008) 17(4) *Environmental Politics* 610.

Bell DA, *East Meets West: Human Rights and Democracy in East Asia* (Princeton University Press 2000).
Bell D, 'Environmental Refugees: What Rights? Which Duties?' (2004) 10(2) *Res Publica* 135.
Bell D, 'Carbon Justice? The Case Against a Universal Right to Equal Carbon Emissions', in S Wilks (ed.) *Seeking Environmental Justice* (Rodolphi 2008).
Bell D, 'Does Anthropogenic Climate Change Violate Human Rights?' (2011) 14(2) *Critical Review of International Social and Political Philosophy* 99.
Bernstein S, et al., 'A Tale of Two Copenhagens: Carbon Markets and Climate Governance' (2010) 39(1) *Journal of International Studies* 161.
Besson S, 'Human Rights: Ethical, Political … or Legal? First Steps in a Legal Theory of Human Rights', in DE Childress (ed.), *The Role of Ethics in International Law* (Cambridge University Press 2012).
Bethlehem D, et al., *The Oxford Handbook on World Trade Law* (Oxford University Press 2009).
Beyerlin U, 'Different Types of Norms in International Environmental Law: Policies, Principles, and Rules', in D Bodansky, J Brunnee and E Hey (eds), *The Oxford Handbook of International Environmental Law* (Oxford University Press 2008).
Beyerlin U and Marauhn T, *International Environmental Law* (Hart Publishing 2011).
Birnie P, Boyle A and Redgwell C, *International Law and the Environment* (3rd edn, Oxford University Press 2009).
Bix B, 'Natural Law: the Modern Tradition', in JL Coleman and S Shapiro (eds), *The Oxford Handbook of Jurisprudence and Philosophy of Law* (Oxford University Press 2002).
Bodansky D, 'The UN Framework Convention on Climate Change: A Commentary' (1993) 18 *Yale Journal of International Law* 451.
Bodansky D, 'Customary (and Not So Customary) International Environmental Law' (1995) 3(1) *Indiana Journal of Global Legal Studies* 105.
Bodansky D, 'The Copenhagen Climate Change Conference – A Post-Mortem' (2010) 104(2) *American Journal of International Law* 230.
Bodansky D (2012a), 'Evaluating Durban' (*Opinio Juris* 2012), opiniojuris.org/2011/12/12/evaluating-durban/ (accessed 26 September 2012).
Bodansky D (2012b), 'The Durban Platform: Issues and Options for a 2015 Agreement' (Center for Climate and Energy Solutions December 2012), http://www.c2es.org/docUploads/durban-platform-issues-and-options.pdf (accessed 19 April 2013).

Bodansky D and Diringer E, 'The Evolution of Multilateral Regimes: Implications for Climate Change' (Pew Center on Global Climate Change 2010).

Bodansky D, Brunnee J and Hey E, *The Oxford Handbook of International Environmental Law* (Oxford University Press 2007).

Boden TA, Marland G and Andres RJ, 'Global, Regional and National Fossil-Fuel CO_2 Emissions' (Carbon Dioxide Information Analysis Center, Oakridge Tennessee 2006), http://cdiac.ornl.gov/trends/emis/overview_2006.html (accessed 12 May 2010).

Borrie J and Martin Randin V, 'A Comparison between Arms Control and Other Multilateral Negotiation Processes', in J Borrie and V Martin Randin (eds) *Alternative Approaches in Multilateral Decision Making: Disarmament as Humanitarian Action* (United Nations Institute for Disarmament Research, Geneva, Switzerland May 2005).

Bosetti V and Frankel J, 'Sustainable Cooperation in Global Climate Policy: Specific Formulas and Emission Targets to Build on Copenhagen and Cancun' (Human Development Research Paper 2011/07) http://hdr.undp.org/en/reports/global/hdr2011/papers/HDRP_2011_07.pdf (accessed 26 September 2012).

Bosselmann K, 'Ecological Justice and Law', in BJ Richardson and S Woods (eds), *Environmental Law for Sustainability: A Reader* (Hart Publishing 2006).

Bosselmann K, *The Principle of Sustainability: Transforming Law and Governance* (Ashgate Publishing 2008).

Bothe M and Rehbinder E (eds), *Climate Change Policy* (Eleven International Publishing 2005).

Bovens L, 'A Lockean Defense of Grandfathering Emission Rights', in DG Arnold (ed.), *The Ethics of Global Climate Change* (Cambridge University Press 2011).

Boyle A, 'Soft Law in International Lawmaking', in M Evans (ed.), *International Law* (2nd edn, Oxford University Press 2006).

Boyle A, 'Between Process and Substance: Sustainable Development in the Jurisprudence of International Courts and Tribunals', in H Christian Bugge and C Voigt (eds), *Sustainable Development in International and National Law* (Europa Law Publishing 2008).

Bretteville C, 'Commentary: The Slowest Sets the Pace' (*Cicerone* 6/2001), www.cicero.uio.no/media/1657.pdf (accessed 12 January 2009).

Broome J, *Counting the Cost of Global Warming* (White Horse 1992).

Brown Weiss E, 'The Planetary Trust: Conservation and Intergenerational Equity' (1984) 11 *Ecology Law Quarterly* 495.

Brown Weiss E, *In Fairness to Future Generations: International Law: Common Patrimony and Intergenerational Equity* (The United Nations University 1989).

Brunnee J, 'Climate Change and Compliance and Enforcement Processes', in R Rayfuse and S Scott (eds) *International Law in the Era of Climate Change* (Edward Elgar 2012).
Brunnee J and Toope SJ, *Legitimacy and Legality in International Law: An Interactional Account* (Cambridge University Press 2010).
Buchanan A, *Justice, Legitimacy, and Self-determination: Moral Foundations for International Law* (Oxford University Press 2004).
Buchanan A and Golove D, 'Philosophy of International Law', in J Coleman and S Shapiro (eds), *Oxford Handbook of Jurisprudence and Philosophy of Law* (Oxford University Press 2002).
Buchner B and Lehmann J, 'Equity Principles to Enhance Effectiveness of Climate Policy: An Economic and Legal Perspective', in M Bothe and E Rehbinder (eds), *Climate Change Policy* (Eleven International Publishing 2005).
Cameron E, 'Human Rights and Climate Change: Moving from an Intrinsic to an Instrumental Approach', (2009–2010) 38 *Georgia Journal of International and Comparative Law* 673.
Campbell T, *Rights, A Critical Introduction* (Routledge 2006).
Caney S, *Justice Beyond Borders: A Global Political Theory* (Oxford University Press 2005).
Caney S, 'Cosmopolitan Justice, Responsibility, and Global Climate Change' (2005) *Leiden Journal of International Law* 747; reprinted in S Gardiner et al. (eds), *Climate Ethics: Essential Readings* (Oxford University Press 2010).
Caney S, 'Cosmopolitan Justice, Rights and Gobal Climate Change' (2006) 19(2) *Canadian Journal of Law and Jurisprudence* 255.
Caney S, 'Climate Change and the Future: Discounting for Time, Wealth and Risk' (2009) 40(2) *Journal of Social Philosophy* 163.
Caney S (2010a), 'Climate Change, Human Rights and Moral Thresholds', in S Humphreys (ed.), *Human Rights and Climate Change* (Cambridge University Press 2009); reprinted in S Gardiner, et al. (eds) *Climate Ethics Essential Readings* (Oxford University Press 2010).
Caney S (2010b), 'Climate Change and the Duties of the Advantaged' (2010) 13(1) *Critical Review of International Social and Political Philosophy* 203.
Caney S, 'Climate Change, Energy Rights, and Equality', in DG Arnold (ed.), *The Ethics of Global Climate Change* (Cambridge University Press 2011).
Carmody C, 'Considering Future Generations – Sustainability in Theory and Practice' (Economic Roundup Issue 3, 2012), http://www.treasury.

gov.au/PublicationsAndMedia/Publications/2012/Economic-Roundup-Issue-3/Report/Considering-future-generations-8212-sustainability-in-theory-and-practice (accessed 26 September 2012).

Childress DE (ed.), *The Role of Ethics in International Law* (Cambridge University Press 2012).

Climate Commission, *The Critical Decade: International Action on Climate Change* (2012) written by T Flannery, R Beal and G Hueston (Climate Commission Secretariat, Department of Climate Change and Energy Efficiency, Commonwealth of Australia), http://climatecommission.gov.au/report/the-critical-decade/ (accessed 15 January 2013).

Cline WR, 'Meeting the Challenge of Global Warming' (paper prepared for the Copenhagen Consensus Challenge programme of the National Environment Assessment Institute, Denmark 2004).

Confalonieri U and Menne B, 'Human Health', in ML Parry, OF Canziani, JP Palutikof, PJ van der Linden and CE Hanson (eds), *Contribution of Working Group II to the Fourth Assessment Report of the Intergovernmental Panel on Climate Change* (Cambridge University Press 2007), http://www.ipcc.ch/pdf/assessment-report/ar4/wg2/ar4-wg2-chapter8.pdf (accessed 30 March 2011).

Cordonier Segger M, 'Sustainable Development in International Law', in H Christian Bugge and C Voigt (eds), *Sustainable Development in International and National Law* (Europa Law Publishing 2008).

Crawford J, 'Democracy and International Law' (1994) 64 *British Yearbook of International Law* 113.

Crocker DA and Linden T (eds), *Ethics of Consumption: The Good Life, Ethics and Global Stewardship* (Rowman and Littlefield Publishers 1998).

Dai X and Diao Z, 'Towards a New World Order for Climate Change: China and the European Union's Leadership Ambitions' in R Wurzel and J Connelly, *The European Union as a Leader in International Climate Change Politics* (Routledge 2010).

D'Amato A, 'Do We Owe a Duty to Future Generations to Preserve the Global Environment?' (1990) 84 *American Journal of International Law* 190.

Des Jardins JR, *Environmental Ethics: An Introduction to Environmental Philosophy* (Wadsworth Publishing Company 2006).

De-Shalit A, *Why Posterity Matters: Environmental Policies and Future Generations* (Routledge 1995).

Dobson A (ed.), *Fairness and Futurity: Essays on Environmental Sustainability and Social Justice* (Oxford University Press 1999).

Donnelly J, 'The Relative Universality of Human Rights' 29 (2007) *Human Rights Quarterly* 281.

Donnelly J, *Universal Human Rights in Theory and Practice* (Cornell University Press 2013).
Doran PT and Kendall Zimmerman M, 'Examining the Scientific Consensus on Climate Change' (2009) 90(3) *EOS, Transactions American Geophysical Union* 22.
Dryzek J, *The Politics of the Earth: Environmental Discourses* (Oxford University Press 1997).
Dryzek J, Downes D and Hunold C, *Green States and Social Movements: Environmentalism in the United States, United Kingdom, Germany, & Norway* (Oxford University Press 2003).
Dryzek J, 'Paradigms and Discourses', in D Bodansky, J Brunnee and E Hey (eds), *The Oxford Handbook of International Environmental Law* (Oxford University Press 2007).
Dryzek J and Stevenson H, 'Global Democracy and Earth System Governance' (2011) 70 *Ecological Economics* 1865.
Dupuy P, 'Formation of Customary International Law and General Principles', in D Bodansky, J Brunnee and E Hey (eds), *The Oxford Handbook of International Environmental Law* (Oxford University Press 2007).
Eckersley R, *Environmentalism and Political Theory: Toward an Ecocentric Approach* (State University of New York Press 1992).
Eckersley R, 'Ecocentrism Explained and Defended', in J Dryzek and D Schlosberg (eds), *Debating the Earth: the Environmental Politics Reader* (Oxford University Press 1998).
Eckersley R, *The Green State: Rethinking Democracy and Sovereignty* (MIT Press 2004).
Eckersley R, 'Moving Forward in the Climate Negotiations: Multilateralism or Minilateralism?' (2012) 12(2) *Global Environmental Politics* 24.
Elliot R, 'The Rights of Future People' (1989) 6(2) *Journal of Applied Philosophy* 159.
FAO, 'The State of the World's Plant Genetic Resources for Food and Agriculture' (1997), http://apps3.fao.org/wiews/docs/swrfull.pdf (accessed 20 January 2009).
Farber DA, 'The Case for Climate Compensation: Justice for Climate Change Victims in A Complex World' (2008) 2 *Utah Law Review* 377.
Feinberg J, 'Duties, Rights and Claims' (1966) 3(2) *American Philosophical Quarterly* 137.
Fitzmaurice M, *Contemporary Issues in International Environmental Law* (Edward Elgar 2009).
Foundation for Democracy and Sustainable Development, 'Examples of parliamentary innovation for sustainable development: Hungary, Finland, Israel … and the UK?' (2009), http://www.fdsd.org/2009/09/

examples-of-parliamentary-innovation-for-sustainable-development-hungary-finland-israel-and-the-uk/ (accessed 28 September 2009).

Franck T, *Fairness in International Law and Institutions* (Clarendon Press 1995).

Freeman M, *Human Rights: An Interdisciplinary Approach* (2nd edn Polity Press 2011).

Gardiner S, 'Ethics and Global Climate Change' (2004) 114 *Ethics* 555–600; reprinted in S Gardiner, et al. (eds), *Climate Ethics: Essential Readings* (Oxford University Press 2010).

Gardiner S, 'Saved by Disaster? Abrupt Climate Change, Political Inertia and the Possibility of an Intergenerational Arms Race' (2009) 20(4) *Journal of Social Philosophy* 140.

Gardiner S, *A Perfect Moral Storm: The Ethical Tragedy of Climate Change* (Oxford University Press 2011).

Gardiner S, et al. (eds), *Climate Ethics: Essential Readings* (Oxford University Press 2010).

Garnaut R (2008a), 'Keynote Address to the "Climate Change and Social Justice Conference"', 3 April 2008 (2008), http://garnautreview.org.au/CA25734E0016A131/WebObj/Transcript_KeynotespeechtoClimateChangeandSocialJusticeConference_RossGarnaut_3April08/$File/Keynote%20speech.pdf (accessed 7 May 2010).

Garnaut R (2008b), 'Transcript of National Press Club Address Launch of Supplementary Draft Report: *Targets and Trajectories* 5 September 2008' (Garnaut Climate Change Review 2008), http://www.garnautreview.org.au/CA25734E0016A131/WebObj/Transcript-NationalPressClub-5September2008/$File/Transcript%20-%20National%20Press%20Club%20-%205%20September%202008.pdf (accessed 19 December 2008).

Garnaut R (2008c), 'The Garnaut Climate Change Review: Final Report, 30 September 2008' (2008), www.garnautreview.org.au-Garnaut (accessed 19 December 2008).

Garnaut R, 'The Garnaut Climate Change Review: Update 2011. Update Paper 5: The Science of Climate Change' (2011), http://www.garnautreview.org.au/update-2011/update-papers/up5-the-science-of-climate-change.html (accessed 27 March 2013).

Gillespie A, *International Environmental Law, Policy and Ethics* (Clarendon Press 1997).

Glenn HP, 'The Ethic of International Law', in Childress DE (ed.), *The Role of Ethics in International Law* (Cambridge University Press 2012).

Goodin RE, *Protecting the Vulnerable: A Reanalysis of Our Social Responsibilities* (The University of Chicago Press 1985).

Gosseries A, 'Cosmopolitan Luck Egalitarianism and the Greenhouse Effect', in D Weinstock (ed.), *Global Justice, Global Institutions*, Volume 31, Canadian Journal of Philosophy Series (University of Calgary Press 2007).

Gramsci A, *Selections from the Prison Notebooks* (International Publishers 1971).

Group of Twenty, 'The Global Plan for Recovery and Reform' (2 April 2009), http://www.g20.org/Documents/final-communique.pdf (accessed 6 May 2010).

Grubb M, 'Seeking Fair Weather: Ethics and the International Debate on Climate Change' (1995) 71(3) *International Affairs*.

Gupta S, et al., 'Policies, Instruments and Cooperative Arrangements', in *Climate Change 2007: Mitigation of Climate Change.* Contribution of Working Group III to the Fourth Assessment report of the Intergovernmental Panel on Climate Change (Cambridge University Press 2007).

Hajer MA, *The Politics of Environmental Discourse: Ecological Modernization and the Policy Process* (Oxford University Press 1995).

Hall S, *Principles of International Law* (LexisNexis Butterworths 2010).

Harris PG, *World Ethics and Climate Change: From International to Global Justice* (Edinburgh University Press 2010).

Hart HLA, *The Concept of Law* (Clarendon Press; Oxford University Press 1994).

Hay P, *Main Current in Western Environmental Thought* (University of New South Wales Press 2002).

Hayward T, 'Constitutional Environmental Rights: A Case for Political Analysis' (2000) 48(3) *Political Studies* 558.

Hayward T, *Constitutional Environmental Rights* (Oxford University Press 2005).

Hayward T, 'Human Rights Versus Emissions Rights: Climate Justice and the Equitable Distribution of Ecological Space' (December 2007) 21(4) *Ethics and International Affairs* 431.

Hey E and Fourie AN, 'Participation in Climate Change Governance and its Implications for International Law', in R Rayfuse and SV Scott (eds), *International Law in the Era of Climate Change* (Edward Elgar 2012).

Hiskes RP, *The Human Right to a Green Future: Environmental Rights and Intergenerational Justice* (Cambridge University Press 2009).

Hoffman AJ, 'Getting Ahead of the Curve: Corporate Strategies that Address Climate Change' (2006) Prepared for the Pew Center on Global Climate Change, available at http://www.pewclimate.org/publications/report/getting-ahead-curve-corporate-strategies-address-climate-change (accessed 19 December 2008).

Höhne N and Blok K, 'Calculating Historical Contributions to Climate Change – Discussing the "Brazilian Proposal"' (2005) 71 *Climatic Change* 141.

Höhne N, et al. (eds), 'Contributions of Individual Countries' Emissions to Climate Change and their Uncertainty' (2011) 106(3) *Climatic Change* 359.

Holland B, 'Justice and the Environment in Nussbaum's "Capabilities Approach": Why Sustainable Ecological Capacity is a Meta-Capacity' (2008) 61(2) *Political Research Quarterly* 319.

Hulme M, *Why We Disagree about Climate Change: Understanding Controversy, Inaction and Opportunity* (Cambridge University Press 2009).

IEA, 'International Energy Agency, World Energy Outlook 2012' (2012), http://www.iea.org/publications/freepublications/publication/English.pdf (accessed 27 March 2013).

International Law Association, 'Legal Principles Relating to Climate Change' (ILA Hague Conference, ILA 2010).

International Law Association, 'Sofia Conference Final Report: International Law on Sustainable Development' (ILA 2012).

IPCC, *Climate Change 1995: Economic and Social Dimensions of Climate Change. Contribution of Working Group III to the Second Assessment Report of the IPCC* (University of Cambridge Press 1996).

IPCC, 'United Nations Intergovernmental Panel on Climate Change, Principles Governing IPCC Work' (2006), http://www.ipcc.ch/pdf/ipcc-principles/ipcc-principles.pdf (accessed 27 March 2013).

IPCC (2007a), 'Climate Change 2007: Synthesis Report: Summary for Policymakers' (2007), http://www.ipcc.ch/pdf/assessment/ar4/syr/ar4.syr _spm.pdf (accessed 17 December 2008).

IPCC (2007b), 'Fourth Assessment Report, Climate Change 2007: Working Group I: The Physical Science Basis' (2007), http://www.ipcc.ch/publications_and_data/publications_ipcc_fourth_assessment_report_wg1_report_the_physical_science_basis.htm.

IPCC (2007c), 'Fourth Assessment Report, Climate Change 2007: Working Group II: Impacts, Adaption and Vulnerability' (2007) 5.8.1, http://www.ipcc.ch/publications_and_data/ar4/wg2/en/ch5s5-8.html#5-8-1 (accessed 31 August 2011).

IPCC (2007d), 'Fourth Assessment Report, Climate Change 2007: Mitigation of Climate Change' (2007) Working Group III Report, http://www.ipcc.ch/ipccreports/ar4-wg3.htm (accessed 10 January 2011).

IPCC (2008), 'Climate Change and Water', Technical Paper VI (2008), BC Bates, ZW Kundzewicz, S Wu and JP Palutikof (eds) Ch 5 http://www.ipcc.ch/pdf/technical-papers/ccw/chapter5.pdf (accessed 10 January 2011).

Ishay M, *The History of Human Rights* (University of California Press 2004).

Jackson T, *Prosperity without Growth: Economics for a Finite Planet* (Earthscan 2009).

Jamieson D, 'Ethics, Public Policy and Global Warming' (1992) 17.2 *Science, Technology, Human Values* 139; reprinted in S Gardiner, et al. (eds), *Climate Ethics: Essential Readings* (Oxford University Press 2010).

Jamieson D, 'Climate Change and Global Environmental Justice', in P Edwards and C Miller (eds), *Changing the Atmosphere: Expert Knowledge and Global Environmental Governance* (MIT Press 2001).

Janicke M, 'German Climate Change Policy: Political and Economic Leadership', in R Wurzel and J Connelly (eds), *The European Union as a Leader in International Climate Change Politics* (Routledge 2010).

Janssen MA, den Elzen MGJ and Rotmans J, 'Allocating CO_2-emissions by using equity rules and optimization' RIVM, Report no. 222901012 (Dutch National Institute for Public Health and the Environment 1992).

Jevrejeva S, Moore J and Grinsted A, 'Sea Level Projections to AD2500 with a New Generation of Climate Change Scenarios' (January 2012) 80–81 *Global and Planetary Change* 14.

Jones CD and Lowe J, 'Committed Ecosystem Changes', in K Richardson, W Steffen and D Liverman (eds), *Climate Change: Global Risks, Challenges and Decisions* (Cambridge University Press 2011).

Jotzo F and Pickering J, 'Climate finance at Doha: what's the damage?' (Development Policy Blog, Development Policy Centre 2012), http://devpolicy.org/climate-finance-at-doha-whats-the-damage-20121212-2/ (accessed 18 April 2013).

Klass A, 'Federalism at Work: Recent Developments in the Public Trust Lawsuits to Limit Greenhouse Gas Emissions' (July 13, 2012, CPR Blog, Centre for Progressive Reform), (2012a) http://www.progressivereform.org/CPRBlog.cfm?idBlog=8092FA68-ADF9-7258-98BF80BAC5FA4 AA7 (accessed 5 September 2012).

Klass A, 'Renewable Energy and the Public Trust Doctrine' (2012b) 45 *UC Davis Law Review* 1021.

Knox J, 'Climate Change in Human Rights Law' (2009) 50 *Virginia Journal of International Law* 163.

Koskenniemi M, *From Apology to Utopia: The Structure of International Legal Argument, Reissue* (Cambridge University Press 2005).

Kotze LJ, 'Phiri, the Plight of the Poor and the Perils of Climate Change: Time to Rethink Environmental and Socio-Economic Rights in South Africa?' (2010) 1(2) *Journal of Human Rights and the Environment* 135.

Kriegler E, et al., 'Imprecise Probability Assessment of Tipping Points in the Climate System' (2009) 106(13) *Proceedings of the National Academy of Sciences* 5041.

Kverndokk S and Rose A, 'Equity and Justice in Global Warming Policy' (2008) No. 21/2008 *Memorandum* 1.

La Viña AGM, 'The Right to a Sound Environment in the Philippines: The Significance of the *Minors Oposa* Case' (1994) 3(4) *Review of European Community and International Environmental Law* 246.

Lange A, Vogt C and Ziegler A, 'On the Importance of Equity in International Climate Policy: An Empirical Analysis' (2007) 29(3) *Energy Economics* 545.

Lange A, et al., 'On the Self-Interested Use of Equity in International Climate Negotiations' (2010) 54(3) *European Economic Review* 359.

Langford M, Vandenhole W, Scheinin M, van Genugten, Chapter 1 Introduction '*An Emerging Field*', in M Langford, W Vandenhole, M Scheinin, and W van Genugten (eds) *Global Justice, State Duties, The Extraterritorial Scope of Economic, Social and Cultural Rights in International Law* (Cambridge University Press 2013), 3.

Langlois AJ, *The Politics of Justice and Human Rights: Southeast Asia and Universalist Theory* (Cambridge University Press 2001).

Lawrence P, 'International Legal Regulation for Protection of the Ozone Layer: Some Problems of Implementation' (1990) 2(1) *Journal of Environmental Law* 17.

Lawrence P, 'The Asia Pacific Partnership on Clean Development and Climate (AP6): A Distraction to the Kyoto Process or a Viable Alternative?' (2007) 10(4) *Asia Pacific Journal of Environmental Law* 183.

Lawrence P, 'APEC Promises a Roar and Delivers a Whimper: The Sydney Declaration on Climate and Energy' (2008) 11(1&2) *Asia Pacific Journal of Environmental Law* 29.

Lawrence P, 'Justice and Future Generations: Environment Discourses, International Law and Climate Change', in B Jessup and K Rubenstein (eds), *Environmental Discourses in International and Public Law* (Cambridge University Press 2012).

Lefeber R, 'Climate Change and State Responsibility', in R Rayfuse and SV Scott (eds), *International Law in the Era of Climate Change* (Edward Elgar 2012).

Lenton TM, et al., 'Tipping Elements in the Earth's Climate System' (2008) 105(6) *Proceedings of the National Academy of Sciences* 1786.

Lesniewska F, 'Filling the Holes: The Montreal Protocol's Non-Compliance Mechanism', in M Fitzmaurice, et al. (eds), *Research Handbook on International Environmental Law* (Edward Elgar 2010).

Levin K and Bradley R, 'Comparability of Annex I Emission Reduction Pledges' (World Resources Institute Working Paper February 2010).

Levy DL and Newell J, 'A Neo-Gramscian Approach to Business in International Environmental Politics: An Interdisciplinary Multilevel Framework', in DL Levy and J Newell (eds), *The Business of Global Environmental Governance* (Massachusetts Institute of Technology 2005) 47.

Light A, 'The Case for a Practical Pluralism', in A Light and H Rolston (eds), *Environmental Ethics: An Anthology* (Blackwell Publishing 2003).

Light A and Pool S, 'The Copenhagen Accord at Three Months. 110 Countries Now Support a New Global Effort to Achieve Climate Safety' (Center for American Progress 29 March 2010), http://www.americanprogress.org/issues/green/news/2010/03/29/7376/the-copenhagen-accord-at-three-months/ (accessed 18 April 2013).

Limon M, 'Human Rights Obligations and Accountability in the Face of Climate Change' (2010) 38(3) *Georgia Journal of International and Comparative Law* 546.

Lomborg B (ed.), *How to Spend $50 Billion to Make the World a Better Place* (Cambridge University Press 2006).

Lomborg B, *Cool IT: The Skeptical Environmentalist's Guide to Global Warming* (Random House 2007).

Lord R, et al., *Climate Change Liability, Transnational Law and Practice* (Cambridge University Press 2012).

Lowe V, 'Sustainable Development and Unsustainable Arguments', in AE Boyle and D Freestone (eds), *International Law and Sustainable Development: Past Achievements and Future Challenges* (Oxford University Press 1999).

Macintosh A, 'Keeping Warming within the 2°C Limit after Copenhagen' (2010) 38(6) *Energy Policy* 2964.

Mackie JL, *Ethics: Inventing Right and Wrong* (Penguin 1977).

Macklin R, 'Can Future Generations Correctly Be Said to Have Rights?', in E Partridge (ed.), *Responsibilities to Future Generations* (Prometheus Books 1981).

MacLean D and Brown PG (eds), *Energy and the Future* (Rowman and Littlefield Publishers 1983).

Maclean I and Wilson RJ, 'Recent Ecological Responses to Climate Change Support Predictions of High Extinction Risk' (2011) 108 (30) *Proceedings of the National Academy of Sciences* 12337–12342, www.pnas. org/content/108/30/1237.shorts (accessed 26 September 2012).

MacNeil R and Paterson M, 'Neoliberal Climate Policy: From Market Fetishism to the Developmental State' (2012) 21(2) *Environmental Politics* 230.

McCrudden C, 'Human Dignity and Judicial Interpretation of Human Rights' (2008) 19(4) *European Journal of International Law* 655.

McGee J and Taplin R, 'The Role of the Asia Pacific Partnership in Discursive Contestation of the Climate Change Regime' (2009) 9(3) *International Environmental Agreements Law, Politics and Economics* 213.

Meadowcraft J, 'Sustainable Development: A New(ish) Idea for a New Century?' (2000) 48 *Political Studies* 370.

Meadows D, et al., *Limits to Growth: The 30 Year Update* (Earthscan 2004).

Meehl GA, Stocker TF, Collins WD, et al., 'Global Climate Projections', in Solomon et al. (eds), *Climate Change 2007: The Physical Science Basis*. Contribution of Working Group I to the Fourth Assessment Report of the Intergovernmental Panel on Climate Change (Cambridge University Press 2007).

Meier AJ, 'Apologies: What Do We Know?' (1998) 8(2) *International Journal of Applied Linguistics* 215.

Meinshausen M, et al., 'Greenhouse-Gas Emission Targets for Limiting Global Warming to 2°C' (2009) 458 *Nature* 1158.

Meyer L, 'Intergenerational Justice', in *Stanford Encyclopedia of Philosophy* (2008), http://plato.stanford.edu/entries/justice-intergenerational/ (accessed 21 August 2009).

Meyer LH and Roser D, 'Climate Justice and Historical Emissions' (2010) 13(1) *Critical Review of International Social and Political Philosophy* 229.

Mill JS, *On Liberty* (1859); reprinted in JS Mill, *Utilitarianism: On Liberty: Essay on Bentham* (World Publishing Company 1962).

Miller D, *Principles of Social Justice* (Harvard University Press 1999).

Miller D, 'Global Justice and Climate Change: How Should Responsibilities Be Distributed?' (The Tanner Lectures on Human Values, delivered Tsinghua University, Beijing, 24–25 March 2008), http://tannerlectures.utah.edu/lectures/documents/Miller_08.pdf (accessed 27 March 2013).

Min SK, Zhang X, et al., 'Human Contribution to More Intensive Precipitation Extremes' (2011) 470 *Nature* 378.

Moellendorf D, *Global Inequality Matters* (Palgrave Macmillan 2009a).

Moellendorf D, 'Treaty Norms and Climate Change Mitigation' (2009b) 23(3) *Ethics and International Affairs* 247.

Moellendorf D, 'Justice and the Assignment of the Intergenerational Costs of Climate Change' (Summer 2009c) 40(2) *Journal of Social Philosophy* 204.

Moellendorf D, 'Common Atmospheric Ownership and Equal Emissions Entitlements', in DG Arnold (ed.), *The Ethics of Global Climate Change* (Cambridge University Press 2011).

Mol APJ and Sonnenfeld DA (eds) 'Ecological Modernization Around the World: An Introduction' (Spring 2000) 9(1) *Environmental Politics* 3.

Muller B, *Justice in Global Warming Negotiations: How to Obtain a Procedurally Fair Compromise* (Oxford Institute for Energy Studies 1999) http://www.oxfordclimatepolicy.org/publications/documents/j2ed.pdf (accessed 3 December 2013).

Neumayer E, 'Global Warming: Discounting is Not the Issue but Substitutability is' (1999) 27(1) *Energy Policy* 33.

Neumayer E, 'In Defence of Historical Accountability for Greenhouse Gas Emissions' (2000) 33 *Ecological Economies* 185.

Neumayer E, 'National Carbon Dioxide Emissions: Geography Matters' (2004) 36(1) *Area* 35.

Newell P, 'The Political Economy of Global Environmental Governance' (2008) 34(3) *Review of International Studies* 507.

Newell P and Paterson M, *Climate Capitalism: Global Warming and the Transformation of the Global Economy* (Cambridge University Press 2010).

Nordhaus W, *A Question of Balance, Weighing the Options on Global Warming Policies* (Yale University Press 2008).

Norton BG, *Toward Unity Among Environmentalists* (Oxford University Press 1991).

Norton BG, 'Environmental Ethics and Weak Anthropocentrism', in A Light and R Holmes (eds), *Environmental Ethics: an Anthology* (Blackwell 2003).

Norton Rose Group, 'Climate Change Commitments in Asia Pacific: A Green Revolution?' (Asia Pacific Climate Policy Series: Japan, Issue 1 May 2011).

Nussbaum MC, *Frontiers of Justice: Disability, Nationality, Species Membership* (Harvard University Press 2006).

Oppenheimer M and Petsonk A, 'Article 2 of the UNFCCC: Historical Origins, Recent Interpretations' (2005) 73(3) *Climate Change* 195.

Oreskes N, 'Beyond the Ivory Tower: The Scientific Consensus on Climate Change' (2004) 306(5702) *Science* 1686.

Oreskes N and Conway EM, *Merchants of Doubt: How a Handful of Scientists Obscured the Truth on Issues from Tobacco Smoke to Global Warming* (Bloomsbury 2010).

Oshitani S, *Global Warming Policy in Japan and Britain: Interactions between Institutions and Issue Characteristics* (Manchester University Press 2006).

Osofsky HM, 'Litigation's Role in the Path of U.S. Federal Climate Change Regulation: Implications of AEP v. Connecticut' (2012) 46(2) *Valparaiso University Law Review*, http://scholar.valpo.edu/cgi/viewcontent.cgi?article=2233&context=vulr (accessed 19 April 2013).

O'Sullivan K, van der Walt A, Kano H and Jasper T, 'Japan's proposed energy policy – how will it affect Australia's energy and resources sector?' (Clayton Utz Insights 27 September 2012), http://www.claytonutz.com.au/publications/edition/27_september_2012/20120927/japans_proposed_energy_policy-how_will_it_affect_australias_energy_and_resources_sector.page (accessed 15 January 2013).

Page E, *Climate Change, Justice and Future Generations* (Edward Elgar 2006).

Page E, 'Intergenerational Justice of What: Welfare, Resources or Capabilities?' (2007) 16(3) *Environmental Politics* 453.

Page E, 'Distributing the Burdens of Climate Change' (2008) 17(4) *Environmental Politics* 556.

Parfit D, *Reasons and Persons* (Oxford University Press 1987).

Partridge E (ed.), *Responsibilities to Future Generations: Environmental Ethics* (Prometheus Books 1981).

Partridge E, 'On the Rights of Future Generations', in D Scherer (ed.), *Upstream/Downstream: Issues in Environmental Ethics* (Temple University Press 1990), http://gadfly.igc.org/papers/orfg.htm (accessed 20 November 2008).

Pearce D, 'The Social Cost of Carbon and its Policy Implications' (2003) 19(3) *Oxford University Press and the Oxford Review of Economic Policy Limited* 362.

Pearse G, *High and Dry, John Howard, Climate Change and the Selling of Australia's Future* (Penguin 2007).

Pearse G, 'Quarry Vision: Coal, Climate Change and the End of the Resources Boom', *Quarterly Essay* 33 (Black Inc 2009).

Pederson O, 'The Janus-Head of Human Rights and Climate Change Adaptation and Mitigation', (2011) 80 *Nordic Journal of International Law* 403.

Peel J and Sands P, *Principles of International Environmental Law* (Cambridge University Press 2012).

Perrings C and Pearce D, 'Threshold Effects and Incentives for the Conservation of Biodiversity' (1994) 4 *Environmental and Resource Economics* 13.

Pettenger ME (ed.), *The Social Construction of Climate Change: Power, Knowledge, Norms, Discourses* (Ashgate 2007).

Phelan L, et al., 'The Political Economy of Addressing the Climate Crisis in the Earth System: Undermining Perverse Resilience' (2012) 18(1) *New Political Economy* 1.

Philibert C, 'International Energy Technology Collaboration and Climate Change Mitigation' (Organisation for Economic Co-operation and Development, International Energy Agency 2004) COM/ENV/EPOC/IEA/SLT(2004)1, www.oecd.org/dataoecd/58/62/32138947.pdf (accessed 25 January 2011).

Pickering J and Barry C, 'On the Concept of Climate Debt: Its Moral and Political Value' (2012) 15(5) *Critical Review of International Social and Political Philosophy* 667.

Pickering J, Vanderheiden S and Miller S, '"If Equity's In, We're Out": Scope for Fairness in the Next Global Climate Agreement' (2012) 26(4) *Ethics and International Affairs* 423.

Pletcher GK, 'The Rights of Future Generations', in E Partridge (ed.), *Responsibilities to Future Generations: Environmental Ethics* (Prometheus Books 1981).

Pogge T, 'A Global Resource Dividend', in DA Crocker and T Linden (eds), *Ethics of Consumption: The Good Life, Ethics and Global Stewardship* (Rowman and Littlefield 1998).

Pogge T, *World Poverty and Human Rights* (2nd edn, Polity 2008).

Posner EA and Sunstein CR, 'Climate Change Justice' (2007) 96 *Georgetown Law Journal* 1565.

Posner EA and Weisbach D, *Climate Change Justice* (Princeton University Press 2010).

Rajamani L, *Differential Treatment in International Law* (Oxford University Press 2006).

Rajamani L, 'The Increasing Currency and Relevance of Rights-Based Perspectives in the International Negotiations on Climate Change' (2010) 22 *Journal of Environmental Law* 3.

Rajamani L, 'The Durban Platform for Enhanced Action and the Future of the Climate Regime' (2012) 61 *International and Comparative Law Quarterly* 501.

Rawls J, *A Theory of Justice* (Harvard University Press 1971).

Rawls J, *Justice as Fairness: A Restatement*, edited by E Kelly (Harvard University Press 2001).

Rayfuse R, 'Climate Change and the Law of the Sea', in R Rayfuse and SV Scott (eds), *International Law in the Era of Climate Change* (Edward Elgar 2012).

Rayfuse R and Scott SV (eds), *International Law in the Era of Climate Change* (Edward Elgar 2012).

Raz J, 'Right-based Moralities', in J Waldron (ed.), *Theories of Rights* (Oxford University Press 1984).

Redgwell C, *Intergenerational Trusts and Environmental Protection* (Manchester University Press 1999).

Regan T and VanDeVeer D (eds), *And Justice for All: New Introductory Essays in Ethics and Public Policy* (Rowman and Littlefield Publishers 1982).

Renteln AD, *International Human Rights, Universalism versus Relativism* (Sage Publications 1990).

Roberts JT and Parks BC, *A Climate of Injustice: Global Inequality, North–South Politics and Climate Policy* (MIT Press 2007).

Robeyns I, 'The Capability Approach', in EN Zalta (ed.) *The Stanford Encyclopedia of Philosophy* (Summer 2011 Edition), http://plato.stanford.edu/archives/sum2011/entries/capability-approach/ (accessed 19 October 2012).

Sagoff M, *The Economy of the Earth* (Cambridge University Press 1988).

Sand P, 'Sovereignty Bounded: Public Trusteeship for Common Pool Resources?' (2004) 4(1) *Global Environmental Politics*, 47–71.

Sand PH, 'The Rise of Public Trusteeship in International Environmental Law', Third International Haub Prize Symposium, Murnau 2013, http://globaltrust.tau.ac.il/the-rise-of-public-trusteeship-in-international-environmental-law/ (accessed 2 September 2013).

Sandel MJ, *Justice: What's the Right Thing to Do?* (Penguin Books 2009).

Sands P and Peel J, *The Precautionary Principle in Practice: Environmental Decision-Making and Scientific Uncertainty* (The Federation Press 2005).

Saul B, et al., *Climate Change in Australia: Warming to the Global Challenge* (Federation Press 2012).

Sceats S and Breslin S, *China and the International Human Rights System* (Chatham House 2012), http://www.chathamhouse.org/sites/default/files/public/Research/International%20Law/r1012_sceatsbreslin.pdf (accessed 23 August 2013).

Schlosberg D and Ahern RJ, Paper prepared for the seminar: *The EU, Climate Change and Global Environmental Governance,* sponsored by the Europa Institute and held at the University of Edinburgh, November 2009), http://dschlosberg.com/Prof_David_Schlosberg/About_Me.html (accessed 17 February 2014).

Scholtz W, 'Equity as the Basis for Future Global Emission Reductions: Between Pragmatic Panacea and Idealistic Impediment. The Optimization of the CBDR Principle via Realism' (2009) 42(2) *The Comparative and International Law Journal of Southern Africa* 166.

Scotford E, 'Mapping the Article 174(2) EC Case Law: A First Step to Analyzing Community Environmental Law Principles' (2008) 8 *Yearbook of European Environmental Law* 1.

Sen A, 'Elements of a Theory of Human Rights' (2004) 32(4) *Philosophy and Public Affairs* 315.

Sen A, *The Idea of Justice* (Belknap Press of Harvard University Press 2009).
Shaw M, *International Law* (6th edn, Cambridge University Press 2008).
Shelton D, 'Equitable Utilisation of the Atmosphere: A Rights-based Approach to Climate Change?', in S Humphreys (ed.), *Human Rights and Climate Change* (Cambridge University Press 2009).
Shue H, *Basic Rights: Subsistence, Affluence and U.S. Foreign Policy* (Princeton University Press 1980).
Shue H, 'Mediating Duties' (July 1988) 98(4) *Ethics* 687.
Shue H, 'Subsistence Emissions and Luxury Emissions' (1993) 15(1) *Law and Policy* 39; reprinted in S Gardiner, et al. (eds), *Climate Ethics: Essential Readings* (Oxford University Press 2010).
Shue H, 'Environmental Change and the Varieties of Justice', in FO Hampson and J Reppy (eds), *Earthly Goods: Environmental Change and Social Justice* (Cornell University Press 1996) 9–29.
Shue H, 'Global Environment and International Inequality' (1999) 75(3) *International Affairs* 531; reprinted in S Gardiner, et al. (eds), *Climate Ethics: Essential Readings* (Oxford University Press 2010).
Shue H, 'Deadly Delays, Saving Opportunities: Creating a More Dangerous World?', in S Gardiner, et al. (eds), *Climate Ethics: Essential Readings* (Oxford University Press 2010).
Shue H, 'Human Rights, Climate Change, and the Trillionth Ton', in DG Arnold (ed.), *The Ethics of Global Climate Change* (Cambridge University Press 2011).
Sikora RI and Barry B (eds), *Obligations to Future Generations* (Temple University Press 1978).
Sim M, 'A Confucian Approach to Human Rights' (2004) 21 (4) *History of Philosophy Quarterly* 337.
Simon JL and Kahn H (eds), *The Resourceful Earth: A Response to 'Global 2000'* (Basil Blackwell 1984).
Singer P, 'One Atmosphere', in P Singer, *One World: The Ethics of Globalization* (Yale University Press 2002); reprinted in S Gardiner, et al. (eds), *Climate Ethics: Essential Readings* (Oxford University Press 2010).
Skott P and Davis L, 'Distributional Biases in the Analysis of Climate Change' (Working Paper, University of Massachusetts Amherst, Department of Economics 2012).
Slade TN, 'Climate Change: The Human Rights Implications, for Small Island Developing Countries' (2007) 37 *Environmental Policy and Law* 215.
Snyder GH and Diesing P, *Conflict among Nations: Bargaining, Decision Making, and System Structure in International Crises* (Princeton University Press 1977).

Soltau F, *Fairness in International Climate Change Law and Policy* (Cambridge University Press 2009).
Splash CL, *Greenhouse Economics: Value and Ethics* (Routledge 2002).
Splash CL, 'The Economics of Climate Change Impacts à la Stern: Novel and Nuanced or Rhetorically Restricted?' (2007) 63 *Ecological Economics* 706.
Steffen W, 'Climate Change 2009 – Faster Change and More Serious Risks' (Australian Government Department of Climate Change 2009).
Stephens T, *International Courts and Environmental Protection* (Cambridge University Press 2009).
Stephens T, 'International Courts and Sustainable Development: Using Old Tools to Shape a New Discourse', in B Jessup and K Rubenstein (eds), *Environmental Discourses in Public and International Law* (Cambridge University Press 2012).
Stephenson P and Boston J, 'Climate Change, Equity and the Relevance of European "Effort-Sharing" for Global Mitigation Efforts' (2010) 10(1) *Climate Policy* 3.
Stern N, *Stern Review: The Economics of Climate Change* (Cambridge University Press 2007); (2006) http://webarchive.nationalarchives.gov.uk/+/http://www.hm-treasury.gov.uk/sternreview_index.htm (accessed 14 February 2009).
Stern N, 'Key Elements of a Global Deal on Climate Change' (London School of Economics 2008), http://www2.lse.ac.uk/GranthamInstitute/publications/Other/Key%20Elements%20of%20a%20Global%20Deal%20-Final%20version%201300%2030-4.pdf.
Stevenson H and Dryzek J, 'The Legitimacy of Multilateral Climate Governance: A Deliberative Democratic Approach' (April 2012) 6(1) *Critical Policy Studies* 1.
Stone CD, 'Ethics and International Environmental Law', in D Bodansky, J Brunnee and E Hey (eds), *The Oxford Handbook of International Environmental Law* (Oxford University Press 2007).
Stott PA, et al., 'Detection and Attribution of Climate Change: A Regional Perspective' (2010) 1(2) *Wiley Interdisciplinary Reviews: Climate Change* 192.
Strauss A, 'Climate Change Litigation: Opening the Door to the International Court of Justice', in WCG Burns and HM Osofsky (eds), *Adjudicating Climate Change: State, National and International Approaches* (Cambridge University Press 2009).
Summary of the Copenhagen Climate Change Conference: 7–19 December 2009, *Earth Negotiations Bulletin*, 12 (459), http://www.iisd.ca/vol12/enb12459e.html (accessed 16 May 2010).
Sykes H (ed.), *Future Justice* (Future Leaders 2010).

Szarka J, 'Climate Challenges, Ecological Modernization and Technological Forcing: Policy Lessons from a Comparative US–EU Analysis' (2012) 12(2) *Global Environmental Politics* 87.

Takao Y, 'The Transformation of Japan's Environmental Policy' (2012) 21(5) *Environmental Politics* 772.

Tellmann SM, 'The Constrained Influence of Discourses: The Case of Norwegian Climate Policy' (2012) 21(5) *Environmental Policy* 734.

Templeman S, 'Intellectual Property' (1998) 1(4) *Journal of International Economic Law* 603.

Tomuschat C, *Human Rights, Between Idealism and Realism* (Oxford University Press 2008).

Traxler M, 'Fair Chore Division for Climate Change' (2002) 28(1) *Social Theory and Practice* 101.

Tremmel JC, *A Theory of Intergenerational Justice* (Earthscan 2009).

Tremmel JC, 'Climate Change and Political Philosophy: Who Owes What to Whom?' (2013) 22 (6) *Environmental Values* 725.

Tremmel JC and Robinson K, *Climate Ethics: Environmental Justice and Climate Change* (I.B. Tauris Publishers forthcoming).

UNDP, Human Development Report Office, 'Fighting Climate Change: Human Solidarity in a Divided World' (2007), http://ideas.repec.org/b/hdr/report/hdr2007-2008.html (accessed 20 January 2011).

UNEP, 'Illegal Trading in Ozone Depleting Substances: Asia and Pacific Region' (2007), http://www.mea-ren.org/files/publications/Illegal%20Trade%20in%20ODS.pdf (accessed 9 March 2013).

UNEP, Ozone Secretariat, 'The 2010 Assessment of the Scientific Assessment Panel' (2010), http://oxone.unep.org/Assessment_Panels/SAP/Scientific_Assessment_2010/index.shtml (accessed 9 March 2013).

UNEP, 'Bridging the Emissions Gap' (Synthesis Report, November 2011) www.unep.org/pdf/unep_bridging_gap.pdf (accessed 5 November 2012).

UNFCCC, 'National Greenhouse Gas Inventory Data for the Period 1990–2010' FCCC/SBI/2012/31 (16 November 2012), http://unfccc.int/resource/docs/2012/sbi/eng/31.pdf (accessed 27 March 2013).

United Nations (2009), Copenhagen Accord, Decision 2/CP15, Report of the Conference of the Parties on Its Fifteenth Session, Copenhagen, 7–19 December 2009, FCCC/CP/2009/11/Add.1 (30 March 2010).

Vanderheiden S, *Atmospheric Justice: A Political Theory of Climate Change* (Oxford University Press 2008).

Vanderheiden S, 'Globalizing Responsibility for Climate Change' (2011) 25(1) *Ethics and International Affairs* 65.

Verschuuren J, *Principles of Environmental Law: The Ideal of Sustainable Development and the Role of Principles of International, European and National Environmental Law* (Nomos 2003).

Victor D, *Climate Change: Debating America's Policy Options* (Council on Foreign Relations 2004).

Voigt C, 'State Responsibility for Climate Change Damages' (2008) 77 *Nordic Journal of International Law* 1.

Waldron J (ed.), *Theories of Rights* (Oxford University Press 1984).

Waldron J, *The Right to Private Property* (Clarendon Press 1990).

Waltz S, 'Universalizing Human Rights: The Role of Small States in the Construction of the Universal Declaration of Human Rights' (2001) 23(1) *Human Rights Quarterly* 44.

Weaver AJ, Zickfield K, Montenegro A and Eby M, 'Long-term Climate Implications of 2050 Emissions Reduction Targets' (2007) 34 *Geophysical Research Letters* 1.

Weitzman ML, 'A Review of *The Stern Review on the Economics of Climate Change*' (2007) XLV *Journal of Economic Literature* 703.

Wendt A, *Social Theory of International Politics* (Cambridge University Press 1999).

Weston BH and Bach T, 'Recalibrating the Law of Humans with the Laws of Nature: Climate Change, Human Rights, and Intergenerational Justice' (Working paper Vermont Law School, the University of Iowa 2008).

WGBU, 'Solving the Climate Dilemma: The Budget Approach' (Special Report from the German Advisory Council on Climate Change 2009).

Winkler H, et al., *Equitable Access to Sustainable Development: Contribution to the Body of Scientific Knowledge* (BASIC Expert Group 2011).

Wissenburg M, 'An Extension of the Rawlsian Savings Principle to Liberal Theories of Justice in General', in A Dobson (ed.), *Fairness and Futurity: Essays on Environmental Sustainability and Social Justice* (Oxford University Press 1999).

World Bank, *The Cost to Developing Countries of Adapting to Climate Change: New Methods and Estimates* (World Bank 2010).

World Commission on Environment and Development, *Our Common Future* (Oxford University Press 1987).

Wright R, *A Short History of Progress* (Carroll & Graf Publishers 2005).

Wurzel RKW and Connelly J, 'The European Union's Leadership Role in International Climate Change Politics Reassessed', in R Wurzel and J Connelly (eds), *The European Union as a Leader in International Climate Change Politics* (Routledge 2010).

Wurzel R and Connelly J, *The European Union as a Leader in International Climate Change Politics* (Routledge 2010).

Young HP and Wolf A, 'Global Warming Negotiations: Does Fairness Matter?' (1991) 10 *Brookings Rev* 46.

Zhang L, Mol APJ and Sonnenfeld DA, 'The Interpretation of Ecological Modernisation in China' (2007) 16(4) *Environmental Politics* 659.

Index